Table of Contents

Acknowledgments

Thanks to all of the people on the Problem Writing Committee for their contributions to AGMT.

And a special thanks to:

Dr. James Li
Mrs. Wendy Li

Dr. Koren Takata
Dr. Ron Takata

Preface

The All Girls Math Tournament was first started as a stand-alone tournament in April 2010 to encourage female participation in math tournaments by holding a math tournament just for girls. It was very well received from the beginning and the Orange County Math Circle had continued the tradition for the next six years, with the newly-formed AGMT organization running it last year.

Unfortunately, the ratio of female to male participants in math competitions has not increased significantly over the years. Many young girls are still intimidated by math and would not even consider entering a math competition. Our aim is to "stem" the tide.

During the Summer of 2016, it was decided to spin off the All Girls Math Tournament into its own independent, student run non-profit organization so that it could focus on expanding the reach of the competitions via both national and international satellites. This venture has been very successful and we are proud to report that AGMT will hold satellites in four countries this year! In addition, AGMT is striving to bring together, as a community, girls who enjoy math by hosting additional events such as All Girls Math Mixers and math classes for girls only. Last year, we started collaborating with local girl scout troops by designing math related activities to aid them in earning STEM badges.

Our main event is the annual All Girls Math Tournament, which is a math competition open to girls in grades 3-8. It started originally in Orange County with about 180 participants in the first tournament. In 2015, we added our first satellite tournament in San Diego. As a result of the success of our first satellite, we have continued to add more over the past few years, including Cypress, Cupertino, and Westminster in California as well as international sites in Toronto (Canada), Shenzhen (China), and Husi (Romania).

The format of the tournament consists of two individual rounds as well as a team activity round and ends with a professional female guest speaker and an awards ceremony. All of our tests are to be taken without a calculator. At the beginning of the tournament, participants are given 40 minutes to solve individually 30 problems in the Sprint Round. Next, they each work on the Target Round, which consists of 8 problems to be solved in 24 minutes. These problems are slightly more difficult compared to the first round. One of our goals is to create an atmosphere that incorporates teamwork so that the participants can "relax" a little and have some fun.

So in the last round, participants in teams of six compete in the Team Round, which consists of some fun activities and challenges. For example, the Team Round has consisted of matchstick puzzles, tangram puzzles, and other fun math games in the past.

Parents have often come up to us and asked where they can get more practice problems for their children. Well, here it is, our very first edition of AGMT's A Great Collection of Math Tournament Problems. Hopefully you will find this book to be helpful as you prepare for future math tournaments and develop your math skills in general. Thank you for your support throughout the years.

Yours Truly,
Natalie Yee
President, All Girls Math Tournament
yeegirl@yahoo.com

Leadership

2018 Leadership

Natalie Yee – President
Grade 11, St. Margaret's Episcopal School

Andrew Vollero – Vice President of Problem Writing
Grade 11, St. Margaret's Episcopal School

Jenna Schindele – Vice President of Operations
Grade 11, University High School

Alexander Kvitsinski – Director of Membership
Grade 11, Laguna Beach High School

Grace Sauers – Director of Marketing
Grade 11, Laguna Beach High School

Lucy Chen – Co-Director of Communications
Grade 10, University High School

Jessica Liu – Co-Director of Communications
Grade 9, Northwood High School

2017 Leadership

Natalie Yee – President, Director of Tournaments
Grade 10, St. Margaret's Episcopal School

Jenna Schindele – Director of Operations
Grade 10, University High School

Andrew Vollero – Director of Problem Writing
Grade 10, St. Margaret's Episcopal School

Alexander Kvitsinski – Director of Membership, Assistant Director of Tournaments
Grade 11, Laguna Beach High School

Sophie Courtney - Director of Marketing
Grade 10, Orange County School of the Arts

Jessica Lee - Director of Communications
Grade 11, University High School

Darcy Chung - Assistant Director of Communications
Grade 9, Sage Hill School

Jerry Li – Assistant Director of Tournaments
Grade 10, University High School

2016 Leadership

Natalie Yee – Director of Tournaments
Grade 9, St. Margaret's Episcopal School

Jason Ye – Director of Tournaments
Grade 11, University High School

Victor Chen – Head Problem Writer
Grade 12, Troy High school

2015 Leadership

Jason Ye –Director of Tournaments
Grade 10, University High School

Victor Chen – Head Problem Writer
Grade 11, Troy High School

2014 Leadership

Michelle Chen –Director of Tournaments
Grade 10, Sage Hill School

Eric Xu – Head Problem Writer
Grade 11, South Torrance High School

Guest Speakers

2017

Ms. Karmyn McKnight

2016

Dr. Wanda Claro

2015

Professor Petra Wilder-Smith

2014

Professor Diane Clemens-Knott

Past Winners

2017 Winners

Individual Winners

3rd Grade:
1. Marissa Huang
2. Sarah Xiong
3. Audrey Huang

4th Grade:
1. Bena Feng
2. Sophia Zhou
3. Allison Hung

5th Grade:
1. Fiona Lee
2. Kristine Lu
3. Sophia Chen

6th Grade:
1. Konnie Duan
2. Aniyah Shen
3. Yi-Ning Li

7th Grade:
1. Audrey Lim
2. Shivana Dhingra
3. Angela Chen

8th Grade:
1. YouBin Cho
2. Selene Huang
3. Sana Manesh

Team Winners

3rd/4th Grade:
1. Allison Hung, Sophia Zhou, Diya Sreedhar, Marissa Huang, Sabina Khizroev, Audrey Huang
2. Katie Karathanasis, Nancy Sun, Sloae Tanenbaum, Kate Kamenstein, Kate Goldberg, Emma Garfin

5th/6th Grade:
1. Kristine Lu, Sophia Chen, Wendy Cao, Tanyi Boga
2. Omir Luqman, Siah Cho, Marisol Guillena, Anne Hiraki, Fiona Lee, Laila Aissi

7th/8th Grade:
1. Parisa Khashayar, Melody Chang, Shrishti Dalal, Samia Sarker, YouBin Cho, Hannah Jang
2. Claire Choi, Elaine Zhou, Selene Huang, Shua Cho, Cindy Zhang

2016 Winners

Individual Winners

3rd Grade:
1. Allison Hung
2. Katelyn Gan
3. Marissa Huang

4th Grade:
1. Kristine Lu
2. Wendy Cao
3. Fiona Lee

5th Grade:
1. Konnie Duan
2. Georgia Goldberg
3. Cassidy Kao

6th Grade:
1. Gina Lee
2. Shivana Dhingra
3. Shrishti Dalal

7th Grade:
1. Hannah Zhang
2. Selene Huang
3. Claire Choi

8th Grade:
1. Darcy Chung
2. Shereen Aissi
3. Jiin Kim

Team Winners

3rd/4th Grade:
1. Wendy Cao, Melody Yu, Kristine Lu, Amy Hwang, Katelyn Gan, Diya Surana
2. Allison Hung, Jessie Chen, Ariana Perez, Marissa Huang, Audrey Huang, Louisa Bouzid

5th/6th Grade:
1. Simone Yang, Elizabeth McMahon, Emily Kim, Lea Park, Colette Reed, Alexandra Keyser
2. Lindsey Yi, Grace Shao, Samantha Shon, Olivia Shon, Virginia Egan

7th/8th Grade:
1. Suhaani Gupta, Soo Hyung Kim, Tasha Nguyen, Mona Reddy Kurra
2. Claire Choi, Selene Huang, Elaine Zhou, Sophie Hwang, Angela Xiang, Hannah Zhang

2015 Winners

Individual Winners

3rd Grade:
1. Yohana Kim
2. Kristine Lu
3. Katherine Given

4th Grade:
1. Konnie Duan
2. Krystal Tran
3. Cindy Ding

5th Grade:
1. Hannah Jang
2. Shrishti Dalal
3. Audrey Lim

6th Grade:
1. Selene Huang
2. Elaine Zhou
3. Claire Choi

7th Grade:
1. Catherine Cui
2. Shivana Anand
3. Sun Jae Kim

8th Grade:
1. Kathy Lee
2. Natalie Yee
3. Sophie Courtney

Team Winners

3rd/4th Grade:

1. Samantha Plageman, Konnie Duan, Chloe Appel, Alexandra Sangster, Cora Baker, and Georgia Goldberg

Tied 2. Claudia Mains, Kempton Gohn, Reaghan Larkin, Kaitlin Tam, Zoe Abarca, and Juhani Pawar

Tied 2. Perla Pingston, Laili Aissi, Manasi Narsina, Yi-Ning Li, and Dominique Oertel

5th/6th Grade:

1. Carly Zhou, Elaine Zhou, Grace Shao, Saaraa Danish, and Kamille Dimayuga
2. Surina Arora, Shivana Dhingra, Sunina Sharma, Claire Choi, Selene Huang, and Sophie Hwang

7th/8th Grade:

1. Natalie Yee, Katherine McPhie, Sophie Courtney, Sun Jae Kim, Sameeha Khan, and Jamie Chang
2. Catherine Cui, Samiksha Komatireddy, Shivana Anand, Audrey Harjanto, Julia Kim, and Dana Lim

2014 Winners

Individual Winners

3rd Grade:
1. Cassidy Kao
2. Kiran Jayasinghe
3. Yohana Kim

4th Grade:
1. Hannah Jang
2. Erin Jeon
3. Audrey Lim

5th Grade:
1. Selene Huang
2. Soohyung Kim
3. Olivia Garrett

6th Grade:
1. Yi-Ann Li
2. Katherine McPhie
3. Jeongyoon Yeo

7th Grade:
1. Kathy Lee
2. Natalie Yee
3. Kelly Tran

8th Grade:
1. Lucy Liu
2. Soumya Ravichandran
3. Kaitlyn Yang

Team Winners

3rd/4th Grade:
1. Vanish Pagdiwala, Ellie Brem, Tessa Baker, Cassidy Kao, Ashley Liu, Uma Joshi

5th/6th Grade:
Yi-Ann Li, Jennifer Frey, Phoebe Wang, Hannah Li, Katherine McPhie

7th/8th Grade:
Jasmine Yang, Frankie Son, Prisha Sadhwani, Kathy Lee, Tiffany Huynh, Sophie Courtney

Foreword

The All Girls Math Tournament events are exciting and fun, and so is this engaging book that includes the materials used in them. The mission of AGMT is to encourage girls to learn mathematics, and mathematics is the basis for both knowledge and practice in science, technology, engineering, computer science, mathematics, statistics, and medicine. So let's ask, why is it important for women and girls to learn science, mathematics, engineering, and medicine? Why should our society work to encourage them to do so? And how can we make it happen?

There are lots of reasons we should get women and girls involved in mathematics. Some reasons are about helping individuals fulfill their full potential. Other reasons are about building a better world by drawing on the talents of all people. I'll say more about these reasons at the end of this foreword. But as for how to do it, a good way is to make it exciting and fun to get involved with mathematics and to bond with others who want to be involved. The All Girls Math Tournaments have lots of math problems that are fun to think about. Here are just a few chosen from this book:

- In 2017, April Fool's Day is on a Saturday. In 2018, what day of the week will April Fool's Day be on? (Assume that there are 365 days in a year.)
- Three animals are arguing about a missing cookie. Frog says that Rabbit ate it, Rabbit says that Frog ate it, and Bird says that Frog is innocent. If exactly 2 animals are lying, who is sure to be innocent?
- If a plap is three plups, and a plup is two ploops, how many ploops does it take to make three plaps?

The problems in the book require only ordinary school mathematics and some problem-solving skills. And the book includes thorough worked-out solutions – not just the individual answers – so that anybody interested in problems of a particular type can learn how to attack all such problems. Anybody – regardless of gender.

Women are half of the human race. But from ancient times to the eighteenth century, the accomplishments of only one female mathematician – Hypatia of Alexandria – are known to historians. Even in the eighteenth and nineteenth centuries, women mathematicians can be counted on the fingers of one hand: Madame du Châtelet, Maria Gaetana Agnesi, Sophie Germain, and Sofia Kovalevskaia. Each of these women, since higher education was just for men, learned mathematics only because of very unusual life circumstances.[1] So as late as the year 1900, the whole female half of the human population had produced only a handful of mathematicians. To put it another way, virtually no mathematicians were women. Social discouragement, coupled with lack of access to higher education, explains why this was so.

My own experience reflects that history. When I went to school, few women were doing mathematics and science. I was the only girl in my high school Math Club. In college I was usually the only woman student in advanced math courses. In my first year of graduate school I was one of only two women students in my 100-student Abstract Algebra class. Though the two of us quickly found each other and worked together, we felt outnumbered and a bit odd from the start. It was tough.

But nowadays in university-level mathematics, there are more and more women, and they feel less alone, less lost, and, above all, less odd. Comparing the 21st century with the year 1900, then, we see what a difference the enlargement of opportunities has made. Today, over 40% of bachelor's degrees in mathematics in the United States are earned by women, and close to 30% of mathematics Ph. D. degrees in the U.S. are now earned by women.[2] We can still do better, but to do so, we need initiatives that encourage young women to learn more mathematics. A student in my History of Mathematics course once wrote, "I was happy to see people with backgrounds like my own contributing to world mathematics." With half the human race being female, surely girls can identify with my student and say, "I want to see women being mathematicians, and I want to see other girls learning mathematics and doing it well."

Now let me return to the question, "Why is it so important to encourage girls and young women to learn mathematics?" First of all, it's exciting to learn these things. Why should guys have all the fun? All young people, female as well as male, have a right to develop their interests and talents. And, as mathematics tournaments and organized groups of students working together demonstrate, mathematics isn't something a person does all

[1]On women in mathematics in general, see the online biographies maintained by Agnes Scott College at https://www.agnesscott.edu/lriddle/women/women.htm. See also the book Women of Mathematics: A Biobibliographic Sourcebook, edited by L. S. Grinstein and P. J. Campbell, Greenwood Press, 1987; and Victor Katz, A History of Mathematics: An Introduction, Pearson, 2009, pp. 189-190, 616-617, 714-715, 787, 874, 896-898, 899. On Kovalevskaia, the first woman Ph. D., I recommend Ann Hibner Koblitz, A Convergence of Lives: Sofia Kovalevskaia, Scientist, Writer, Revolutionary, Birkhaüser, 1983.

[2]These statistics fluctuate a bit from year to year but have been fairly consistent throughout so far in the 21st century. Here is one set of fairly recent data in all STEM (Science, Technology, Engineering, Mathematics) fields:
http://www.catalyst.org/knowledge/women-science-technology-engineering-and-mathematics-stem

alone. Mathematics is a social activity. Working in groups with people who share your interests and experiences helps everybody do better and enjoy what they're doing.

But encouraging girls and women to do mathematics does not just benefit them as individuals. Modern society needs capable scientists, mathematicians, and engineers. The larger the pool of interested and talented people we draw from, the more outstanding people we will find. Our world is awash in mathematics and its applications: computers, public opinion polls, business, taxes, credit cards, sports statistics, music and ways of making music, art and architecture, medicine and health – the list could go on forever. Everyone needs to understand a fair amount of mathematics, both for self-preservation in the economy and for effective citizenship. I applaud the organizers of AGMT for their contribution to involving girls in mathematics. I hope that this book finds many enthusiastic readers, and I am sure that they as individuals, and indeed society at large, will benefit from its contents.

Judith V. Grabiner
Flora Sanborn Pitzer Professor of Mathematics Emerita
Pitzer College, Claremont, California 91711

How to Take the Contest

Rules, Timing, etc.

---

1. When taking these practice contests, it is recommended to not use a calculator. Calculators are not allowed on any AGMT contest, and every problem can be solved without one.

2. In the official AGMT, participants are given 40 minutes to complete the Sprint Round, and 24 minutes to complete the Target Round.

3. Please leave all answers in exact form. For example, do not round π to 3.14 or $\frac{22}{7}$, and do not round $\sqrt{2}$ to 1.4. Leave as π and $\sqrt{2}$, respectively.

4. Please put all fraction answers in lowest terms. For example, if the answer to a question is $\frac{1}{2}$, then $\frac{2}{4}$ would be marked as incorrect.

5. Units are not required for an answer to be correct, but they must be correct if they are provided. For example, if the answer to a question is 16 inches, 16 or 16 inches would be marked as correct, but 16 feet would be marked incorrect.

6. Most importantly, do NOT be discouraged if you cannot solve a problem. Many of these problems are meant to be difficult. If you are unsure how to solve a problem, skip it and come back to it later. If you still do not know how to solve it, refer to the solution guide after the test is over.

7. If you have any extra questions, comments, or concerns regarding the problems or the book in general, feel free to email us at: allgirlsmathtournament@gmail.com.

Contests

3rd-4th Grade 2014

Sprint Round

---

1. How much does Alice pay if she buys 42 computers for $700 each?
2. What is $\frac{2}{9} \div \frac{8}{3}$? Express your answer as a common fraction.
3. A baby is crying at a rate of 2 tears per second. How many tears come out after 3 minutes of crying?
4. John is buying cookies. If they cost 35 cents per cookie, how much (in dollars) do 11 cookies cost? Express your answer as a decimal rounded to the nearest hundredth.
5. A wave of minions is eating bread. If they finish one loaf of bread in 8 seconds, how many minutes does it take for them to finish 45 loaves of bread?
6. What is $9 - 8 + 7 - 6 + 5 - 4 + 3 - 2 + 1$?
7. If 1400 cushions are as heavy as 3500 pillows, then how many cushions are as heavy as 5000 pillows?
8. Jamie has a .554 batting average. Allison has a .336 batting average. What is Jamie's average minus Allison's average? Express your answer as a decimal rounded to the nearest thousandth.
9. What is the remainder when $820 \div 17$?
10. What time is it 3000 seconds after 1:30?
11. Helium has 2 protons, Beryllium has 4 protons, Barium has 56 protons, and Mercury has 80 protons. If I have two Heliums, one Beryllium, one Barium, and four Mercuries, how many protons do I have?
12. How many positive factors does the number 14 have?
13. Jamie is cooking 3 pounds of chicken. If he has to wait 35 minutes for every pound of chicken to cook, how many minutes does he have to wait for all of his chicken to finish cooking?
14. Suppose John wants to cut his huge ruler, which is 144 inches long, into 12 different rulers, which are each 12 inches long. How many cuts does he make in his huge ruler?
15. If an ice cube tray is made up of a 3×9 grid of squares. Each square can contain one ice cube. How many ice cubes can be made from 3 trays?

16. The Problem Writing Committee writes 10 problems in 1 minute. How many seconds does it take for the committee to write 16 problems?

17. Sarah's house has 5 stories. If the first story has 10 windows, and the following stories have one less window than the story below it, how many windows does Sarah's house have?

18. Mary keeps all of her money in a piggy bank. On Monday, she takes out one dollar. On Tuesday, she puts in two dollars. On Wednesday, she removes four dollars. On Thursday, she puts in eight dollars. If she has fifty dollars in the bank on Friday, how many dollars did she have before Monday?

19. The mad scientist Eric is testing his new invention, the Best Enlarging Ray Gun, or BERG. The BERG enlarges objects by a factor of 8 in all dimensions. If Eric fires the BERG at a popsicle that is originally 3.5 inches long, what is the final length of the popsicle (in inches)?

20. If x is the greatest prime factor of 36 and y is the greatest prime factor of 27 then what is $x + y$?

21. A certain poodle usually drinks water at a rate of 1 liter per 7 minutes. However, since the poodle is very thirsty today, its drinking rate is tripled. If we leave six-sevenths of a liter out for the poodle, in how many minutes will it finish?

22. What is the value of $5\% + 0.05 + \frac{5}{100}$? Express your answer as a decimal rounded to the nearest hundredth.

23. Olivia likes to play basketball every other day of the week. If she starts playing on a Tuesday, how many times will she play basketball in two weeks? (The two weeks begin on Tuesday and end on Monday 13 days later.)

24. When a turtle wants to cross a river, he has to pay $4. If he crosses the river 5 times on Monday, how much did he pay on Monday?

25. Serge has a crush on Pam and decides to buy her 5 yards of yarn (and some roses). However, the yarn can only be bought in inches. How many inches of yarn does he need to buy?

26. Allison took a series of tests during her freshman year. Her test scores were 92, 85, 87 and 100. However, instead of 100, her test should have been a 98.5. What is the new median? Express your answer as a decimal rounded to the nearest tenth.

27. A dog runs toward a ball 30 feet away. If the dog is halfway there, how much farther does he have to run to get the ball and get back to his starting point (in feet)?

28. If the beaver slaps his tail on water 4 times in 5 seconds, then pauses for 5 seconds, then repeats the cycle, how many times does he slap his tail on the water in one minute?

29. If the lengths of two of the sides of an isosceles triangle are 6 and 12, then what is the perimeter of this triangle?

30. If there are 300 cubbyholes and Joe has 301 flyers, what is the probability that one cubby will have at least two flyers?

Target Round

1. On the planet Tums, someone has a stomach ache every 30 seconds. How many stomach aches occur in 42 minutes?
2. I have four nickels, five quarters, and 6 dollar bills. How much money do I have left if I spend $1.50 on lollipops? Express your answer as a decimal rounded to the nearest hundredth.
3. How much wood (in logs) could a woodchuck chuck (in 35 minutes) if a woodchuck could chuck wood (at a rate of 36 logs per hour)?
4. Harry receives a $50 parking ticket for parking his broomstick in a red zone. Because he's Harry, he gets a 20% discount. How much does Harry have to pay (in dollars)?
5. Danielle is buying lunch at school. She has a choice of a sandwich, popcorn chicken, or a burger plus one of 4 drinks, one of 3 fruits, and one of 5 vegetables. How many choices does she have for her lunch if she gets one of each type?
6. Sarah is delivering newspapers on a street of houses numbered 1 to 100 with even-numbered houses on the right and odd-numbered houses on the left. If she only delivers to house numbers that are a multiple of 3 and on the right, how many houses does she deliver newspapers to?
7. A room is one-half full of people. After 20 people leave, the room is one-third full. How many people would be in the room if it were filled?
8. Paul is not very good at talking to girls. Every time Paul tries to talk to a girl, he gets very nervous and starts counting in multiples of 3. If Paul counts in multiples of 3 (3, 6, 9, 12, 15,...) until he reaches 1,242, how many numbers has Paul said?

3rd-4th Grade 2014 Answer Key

Sprint Round

1. \$29,400
2. $\frac{1}{12}$
3. 360 tears
4. \$3.85
5. 6 minutes
6. 5
7. 2,000 cushions
8. 0.218
9. 4
10. 2:20
11. 384 protons
12. 4 factors
13. 105 minutes
14. 11 cuts
15. 81 ice cubes
16. 96 seconds
17. 40 windows
18. 45 dollars
19. 28 inches
20. 6
21. 2 minutes
22. 0.15
23. 7 times
24. \$20

25. 180 inches
26. 89.5
27. 45 feet
28. 24 times
29. 30
30. 100% or 1

Target Round

1. 84 stomach aches
2. $5.95
3. 21 logs
4. $40
5. 180 choices
6. 16 houses
7. 120 people
8. 414 numbers

3rd-4th Grade 2015

Sprint Round

---

1. What is $2015 \times 20 \times 15 \times 2 \times 0 \times 1 \times 5$?
2. How many feet is 26 yards?
3. If I have 205 pennies and exchange them for some number of nickels, how many nickels would I have?
4. It is currently 2:08 PM. How many minutes will pass before it is 4:01 PM (of the same day)?
5. The area of a square is 49 square feet. What is the length of the side of the square (in feet)?
6. The perimeter of an equilateral triangle is the same as that of a square. If the side length of the square is 12, what is the side length of the triangle?
7. If $a||b = \frac{a+b}{ab}$, what is $4 \mid\mid (5 \mid\mid 6)$? Express your answer as a common fraction.
8. Matthew Tang and Arthur Liu are playing a card game in their free time. If each card game takes 160 seconds, how many games are they going to play in 2 hours?
9. If there are 5 quarks in a quack, 3 quacks in a quat, and 7 quats in a quirt, how many quarks are in 20 quirts?
10. A 23-page packet contains a total of 3450 words. On average, how many words are on each page?
11. A person walks into the airport every 5 seconds. How many people walk into the airport in 20 minutes?
12. What is the least number of coins I need to make 37 cents?
13. Person A, Person B, and Person C have 54 dollars each. Person A uses $\frac{1}{2}$ of their money, Person B uses $\frac{2}{3}$ of their money, and Person C uses $\frac{2}{9}$ of their money to buy food. How many dollars do they have left altogether?
14. Compute the sum of the first 11 even natural numbers.
15. What is the value of $7 \times (2 + 3 - 4 \times 2)$?

16. Train A leaves New York and travels north at 80 mph. Train B leaves New York at the same time but travels south at 100 mph. After 3 hours, how many miles apart are the two trains?

17. If today is Saturday, what day was it 100 days ago?

18. What is the probability that a randomly chosen year in the interval 2000 to 2100 (inclusive) will be a leap year? Express your answer in a common fraction. (Leap years are those that are a multiple of 400 or a multiple of 4 but not 100.)

19. You have a rectangle, and if you increase the length by 10% and increase the width by 5%, by what percent is the area increased by? Express your answer as a decimal rounded to the nearest tenth.

20. Bill has 5 unique Rubik's Cubes to give away to 4 people: Leon, Suhn, Mitchell, and Bill 2.0. How many ways can he distribute them?

21. What is the positive difference between $1 + 2 + 3 + 4 + 5 + \ldots + 100$ and $2 + 3 + 4 + 5 + 6 + \ldots + 101$?

22. A sequence is defined by $a_n = a_{n-1} \times n$. If a_5 is equal to 180, find the value of a_2.

23. The sides of a right triangle have lengths 8 cm, x cm, and 15 cm, and x is an integer. What is the area of the triangle (in square centimeters)?

24. In triangle ABC, angle A is equal to 65 and angle B is equal to 75. The bisector of angle C is drawn to meet AB at point D. What is the measure of angle ADC? (An angle bisector is a line that cuts an angle into two angles of equal measure.)

25. Five friends, Natalie, Lucy, Amy, Thomas, and Ian, go to Disneyland together. They each buy their own ticket for $156 each. They equally split up the lunch bill of $100 and they each buy one pair of Mickey Mouse ears for $22 a pair. If this is all they spent, how many dollars did each of them spend at Disneyland?

26. William, Zack, Bryan, Jason, and Joey are running in a race. These statements are true about the results:

    William finished before Jason but after Joey.

    Jason finished in third place.

    Bryan finished after Joey but he did not finish last.

    Zack finished after both Bryan and Jason.

    Who was the first place runner?

27. If the lengths of two sides of a isosceles triangle are 5 and 8 respectively, what is the sum of all the possible perimeters for the triangle?

28. Bob sells gum in packs of 3 and 5. What is the largest amount of gum that a person is not able to buy using a combination of Bob's packs of gum? For example, a person can get 14 pieces of gum by ordering 3 packs of 3 and 1 pack of 5.

29. Susan is a criminal case solver. Today, her case is to find out who ate which type of candy, Skittles, M&M's, Jolly Ranchers, and Twix. She knows Michael hates chocolate. Mindy says she would never eat Twix in a million years. John said that he saw a boy eat Skittles. Lastly, Joey is under a lot of pressure, and confesses that he ate the Jolly Ranchers. Given that nobody ate the same candy, what did Mindy eat? (Assume M&M's and Twix are the only two chocolate candies.)

30. Using a standard deck of cards, a class is doing a fun game. When a certain suit is a pulled from the deck, the class has to do a certain exercise. Hearts are jumping jacks, diamonds are sit-ups, clubs are push-ups, and spades are burpies. If all face cards (including aces) represent 10, and the number on the card represent how many of each exercise you need to do, how many sit-ups and push-ups would a student have completed after going through the whole deck?

Target Round

1. Compute $\frac{25\times7+2}{3}$.
2. Jenna has 10 pieces of candy. She gives seven pieces to her brother. Then her friend gives her 2 pieces of candy. How many pieces of candy does she have now?
3. How many cents is 26 pennies, 3 nickels, and 5 dimes?
4. What is the area of a rectangle with side lengths 5 and 7?
5. What is the result when the fifth even natural number is multiplied by the fifth odd natural number?
6. A strange clock chimes 1 time at 1 o' clock, 2 times at 2 o'clock, 3 times at 3 o' clock, 4 times at 4 o' clock, 5 times at 5 o' clock, and so on. How many times will the clock chime from 1-12 o' clock?
7. The perimeter of a square is 5 more than the perimeter of a triangle with side lengths 4, 5, and 6. What is the area of the square?
8. How many integers are there between 1 and 100 (inclusive) that can be expressed as the sum of three positive consecutive integers?

3rd-4th Grade 2015 Answer Key

Sprint Round

1. 0
2. 78 feet
3. 41 nickels
4. 113 minutes
5. 7 ft
6. 16
7. $\frac{131}{44}$
8. 45 games
9. 2,100 quarks
10. 150 words per page
11. 240 people
12. 4 coins
13. $87
14. 132
15. -21
16. 540 miles
17. Thursday
18. $\frac{25}{101}$
19. 15.5%
20. 1024 ways
21. 100
22. 4
23. 60 square cm
24. 95°

25. \$198

26. Joey

27. 39

28. 7 pieces of gum

29. M&Ms

30. 188

Target Round

1. 59

2. 5 pieces of candy

3. 91 cents

4. 35

5. 90

6. 78 times

7. 25

8. 32 integers

3rd-4th Grade 2016

Sprint Round

---

1. You had 20 apples. You ate half of your apples. Your friend gave you 6 more apples. How many apples do you have now?

2. What is the area of a triangle with height 7 and base 8?

3. The perimeter of a square with side length 6 is twice the perimeter of an equilateral triangle. What is the side length of the triangle?

4. Maya thinks of a number, multiplies it by 7, adds 14, and says the result out loud. If Maya says "84", what number was she originally thinking of?

5. Clarisse flips a coin twice. What is the probability that the two flips yield different results?

6. A regular pentagon has one side with the value 6. What is the total perimeter?

7. Suppose that $a \diamond b = \frac{a \times b}{a+b}$. Compute $2 \diamond (3 \diamond 6)$.

8. Find the number of divisors of 24, including 1 and 24.

9. Thirteen 2 by 2 squares are pasted together in a straight line. What is the perimeter of the resulting figure?

10. Allison has to solve 15 math problems for homework. 5 are easy problems, 5 are medium problems, and 5 are hard problems. If it takes her 2 minute to solve an easy problem, 4 minutes to solve a medium problem, and 7 minutes to solve a hard problem, how many minutes does it take her to finish her homework?

11. Kate is eating a square pizza. Her sister, Rachel, is eating a rectangular pizza with side lengths 4 and 9. If they both end up eating the same amount of pizza, what is the side length of the square pizza?

12. Hannah pastes together four identical rectangles of length 4 and width 7 to form one large rectangle. What is the area of the final rectangle?

13. A school has some books. All the books can be split evenly among 6 classrooms. They can also be split evenly among 10 classrooms. What is the smallest number of books that the school can have?

14. Ara uses a keyboard with a total of 104 keys. She cannot read yet, so she randomly presses two keys, one after another. What is the probability that both keys are letters? (There are 26 letters in the alphabet).

15. A square and an equilateral triangle have side length 5. Michelle glues one edge of the equilateral triangle to one edge of the square, keeping the square and the triangle touching at only one edge. What is the perimeter of this new figure?

16. A triangle has two angles of 20° and 50°. What is the measure of the third angle?

17. A lake is filled with 20 fish. After every morning, 3 fish swim out of the lake and after every night, 2 fish swim in. After how many mornings will the lake first be empty?

18. Angle A in triangle ABC has an angle measure of 42 degrees. What is the average angle measure of the other two angles of ABC?

19. Two apples weigh as much as three oranges. Nine oranges weigh as much as ten pears. How many apples weigh as much as 40 pears?

20. $3x - 4y = 18$ and $4x - 3y = 10$. Compute $x - y$.

21. Two chipmunks can build a treehouse in five hours. How much time will it take for ten chipmunks to build twenty treehouses?

22. How many squares of side length 2 can fit inside a 5 by 7 rectangle with sides parallel to the sides of the rectangle and no overlap allowed?

23. A square frame for a square painting is made with a thickness of 1 inch. If the area of the frame is three times the area of the painting, what is the side length, in inches, of the painting?

24. Julia, Joey, and Jackie are going to Disneyland together. They each pay \$158 for their ticket. When they get there, they spend \$60 on lunch collectively and \$72 on dinner collectively. If they split the bill equally for both meals and add their ticket price to their total, what did each person pay in total?

25. Isabella has a bag of 15 green, blue, and violet marbles. There are 9 blue marbles, and twice as many green marbles as violet marbles. If she randomly picks one marble, and without putting it back, randomly picks another marble, what is the probability that both marbles are green?

26. Jack has some two-liter water bottles and some three-liter water bottles. He wants to fill a pail of 17 liters while wasting no water. What is the least number of two-liter bottles that Jack has to use?

27. Ilene chose a two digit number. She then reversed the digits of the number: for example, from 19 to 91. She finally added these two numbers together and got 143. What is the sum of the digits of her original number?

28. A rectangle with integer side lengths has an area of 42. How many different possible values are there for its longer side?

29. Jason picks a positive integer. He multiplies his number by itself. Then, he subtracts 1. He gets the answer 399. What was Jason's original number?

30. Mohab is made up of a circle for his head, a square for his body, and 4 congruent rectangles for his arms and legs with no overlapping regions. The circle has a radius that is equivalent to each side of the square, 4 units. The dimensions of each rectangle are 6 units by 1 unit. What is Mohab's overall area? Leave your answer in terms of π.

Target Round

---

1. How many edges does a rectangular prism have?
2. Jeff the sea lion is shaped awfully like a triangle. In fact, he is a triangle. Jeff is four feet wide and eight feet tall! What is half of his area?
3. The perimeter of a regular pentagon is equivalent to the perimeter of an equilateral triangle. If the side length of the pentagon is 6, what is the side length of the triangle?
4. Bessie has a total of 7 animals who are either cows or chickens. All of her animals have a total of 22 legs. How many of Bessie's animals are cows?
5. At Karishma's Restaurant, a meal consists of one appetizer, one main dish, and one drink. There are 5 choices for an appetizer, 10 choices for a main dish, and 9 choices for a drink. If Cathy is allergic to 2 of the appetizers and 5 of the main dishes, how many choices does she have for a meal?
6. Jan invited her friends to her house for a party. At 9 PM, half of the people leave. At 10 PM, one-third of the remaining people leave. Finally, at 11 PM, three-fourths of the remaining people leave, leaving only Jan. How many people, including Jan, were at the party?
7. At a certain store, 5 books and 1 pencil cost twice as much as 2 books and 2 pencils. 1 pencil usually costs 1 dollar. On a special savings day, 4 pencils are given for every purchase of a book. If Patricia needs at least 4 books and 30 pencils, what is the least amount of money she can spend?
8. How many different rectangles with integer side lengths have an area of 100? Two rectangles are considered the same if one can be rotated to get the other one.

3rd-4th Grade 2016 Answer Key

Sprint Round

1. 16 apples
2. 28
3. 4
4. 10
5. $\frac{1}{2}$
6. 30
7. 1
8. 8 divisors
9. 56
10. 65 minutes
11. 6
12. 112
13. 30 books
14. $\frac{1}{16}$
15. 25
16. $110°$
17. 18 mornings
18. $69°$
19. 24 apples
20. 4
21. 20 hours
22. 6 squares
23. 2 inches
24. $202

25. $\frac{2}{35}$

26. 1 two-liter bottle

27. 13

28. 4 possible values

29. 20

30. $40+16\pi$

Target Round

1. 12 edges
2. 8 feet^2
3. 10
4. 4 cows
5. 135 choices
6. 12 people
7. $23
8. 5 rectangles

3rd-4th Grade 2017

Sprint Round

---

1. To participate in the All Girls Math Competition, Casey needs to buy some pencils and erasers. A pencil costs 10 cents and an eraser costs 5 cents. If Casey wants 2 pencils and 2 erasers, how much money would Casey need, in cents?

2. Susie went to the movies and brought twenty dollars. Her friends doubled the amount of money she had. Then, she spent \$6 on popcorn and \$3 on a soda. How much money did Susie have left?

3. Sally and Susie are comparing heights. Sally is 3 feet and 11 inches, while Susie is 5 feet and 1 inch. In inches, what is the difference between Sally's height and Susie's height?

4. Jenna was making a birthday cake, so she buys 2 dozen eggs. If she had 5 eggs left over, how many eggs did she use to make the cake?

5. Tyler gives Emily a \$250 Verizon gift card. If Verizon charges a \$50 one-time fee to get a phone number, and then charges \$40 per month for a phone plan, how many full months can the gift card buy if Emily does not have a phone number?

6. Cheryl receives a KFC coupon in the mail. The coupon gives 50% off a bucket of fried chicken. If a bucket of fried chicken normally costs \$5.00, how much money will the bucket of chicken cost with the coupon?

7. If Iris rolls a six-sided die, what is the probability that she rolls a six? (Please answer using a fraction.)

8. Jackie has a bunch of identical square Lego pieces, so she makes them into a rectangle that is 6 pieces wide and 8 pieces long. Then, using the same Lego pieces, she made a rectangle that was 4 pieces wide. How many pieces long is the new rectangle?

9. Prom King and Queen are being decided at the All Girls Math Tournament Annual Dance. Jane receives 5 more votes than Karen's amount of votes tripled. If Jane received 23 votes, then how many votes did Karen receive?

10. For the All Girls Tournament, Drew has to write 30 problems for the Sprint Round and 8 problems for the Target Round. If it takes 5 minutes to write a Sprint Round Problem, and 10 minutes to write a Target Round Problem, how many minutes does it take for Drew to finish writing all of his problems?

11. One day, Ms. Chou decided to measure the height of her nine students. Their heights are 49 in, 52 in, 52 in, 52 in, 53 in, 53 in, 54 in, 54 in, and 56 in. What is the positive difference between the median height and the mode height, in inches?

12. How many lines of symmetry does the following figure have?

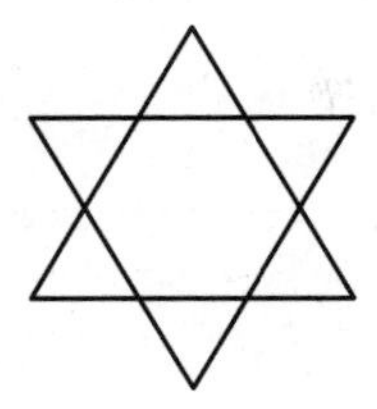

13. Kourtney goes shopping at Justice. A shirt there costs \$10, and a skirt costs \$12. If Kourtney has exactly enough money to buy 6 shirts, how many skirts could she buy?

14. Nessa wants to tile a room that is 5 feet long and 5 feet wide. If the tiles are each 15 inches long and 15 inches wide, how many tiles does she need to tile the floor of the room?

15. If $3 \# 1 = \frac{3*1}{3-1}$, then what is $6 \# 4$?

16. A cosmetic product, called Barbie's Amazing Lip Gloss, either works or does not work. Chloe buys 20 of the lip gloss containers, and she soon discovers that 16 containers work but 4 do not. If the 20 lip gloss containers are put in a bag, and Chloe chooses one out of the bag at random, what is the probability she grabs a lip gloss that does NOT work? (Please answer using a fraction in lowest terms.)

17. All of the small triangles in the figure below are congruent. If one of the triangles has an area of 4, what is the area of the entire figure?

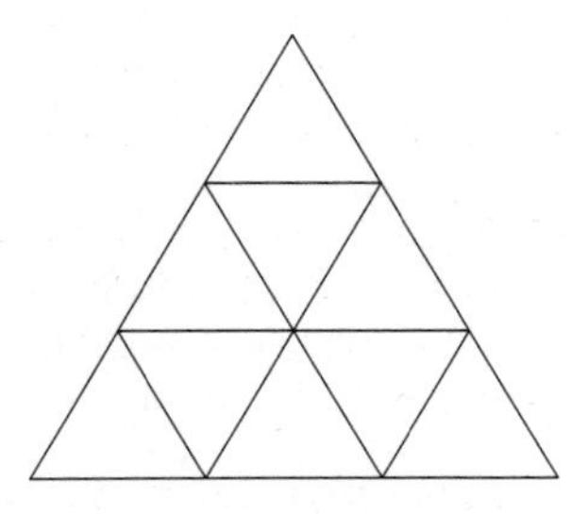

18. In 2017, April Fool's is on a Saturday. In 2018, what day of the week will April Fools be on? (There are 365 days in a year.)

19. Mrs. Runge has a certain number of desks in her classroom. If she arranges the desks into rows of 3, she will have 2 desks that are not in a row. If she arranges the desks into rows of 5, she will have 4 desks that are not in a row. If she has less than 25 desks, how many desks does she have?

20. What is the ones digit of 5×372?

21. A certain rectangle is 4 inches long and 3 inches wide. If the length of the rectangle is doubled, and the width of the rectangle is tripled, what is the area of the new rectangle?

22. Joe has a house of cards that is built with cards stacked like the figure below. If he makes a card house that has 4 stories, how many cards does he need to put in it?

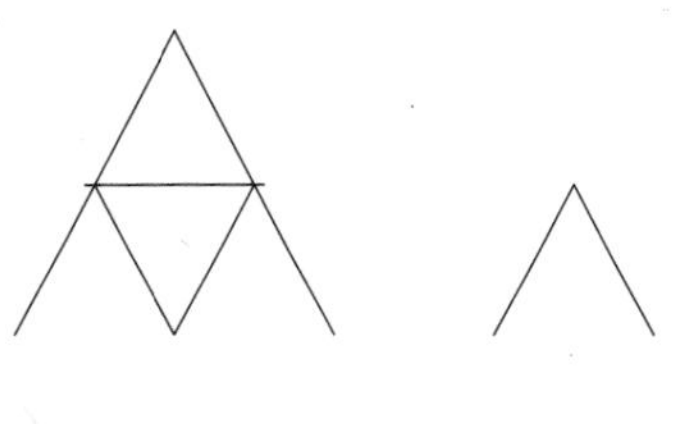

2 story card house 1 story card house

23. Billy Jean draws 5 squares side by side, as shown in the figure below. If the squares have side lengths of 1, 2, 3, 4, and 5, in that order, what is the area of the figure?

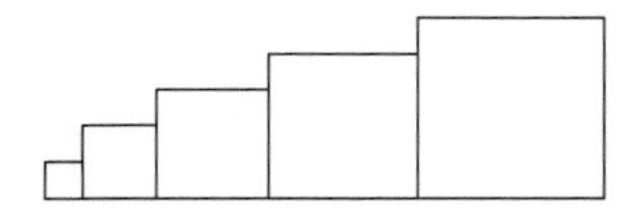

24. What is the perimeter of Billy Jean's figure in the previous problem?

25. Lara and Natasha had 120 marbles in total. After Lara gave Natasha 30 marbles, Natasha had twice as many marbles as Lara. How many marbles did Natasha have *before the exchange*?

26. Carla is planning to use her allowance of \$20 every week in order to buy a new phone, which costs \$200. However, the price of the phone rises by 50%, but luckily she gets a 20% discount on the new price. How many weeks will it take her to save up enough money to buy the phone?

27. If a plap is three plups, and a plup is two ploops, how many ploops does it take to make three plaps?

28. Jenna has a square that is six inches long on each side. She also has 4 triangles that have six-inch bases, and are 4 inches tall. If she glues one triangle to each side of the square, what is the area of the figure?

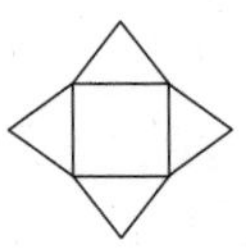

29. If four people working for four hours can complete eight jobs, then how many jobs can eight people complete in eight hours?

30. Miki is doing laps around her school's football field, which is 320 feet long and 160 feet wide. If it takes her 20 seconds to run the width of the football field, how many seconds does it take for her to run a complete lap around her school's football field?

Target Round

1. Kevin really loves reading the Magic Tree House series, and he reads it every day. On weekdays, he reads 25 pages a day, and on weekends, he reads 45 pages a day. If a Magic Treehouse book is 150 pages long, how many complete books can Kevin finish in five weeks?

2. How many triangles are in the figure below? (Hint: There are more than five!)

3. Chyna wanted to help with her school's bake sale, so she decided to make some cookies. In Chyna's cookie recipe, she needs $1\frac{1}{2}$ cups of flour to make 12 cookies. If she wants to make 72 cookies, how many cups of flour does she need?

4. There are about 7 billion people in the world, and about $\frac{3}{5}$ of them live in Asia. If $\frac{2}{3}$ of Asians do NOT live in China, about how many people live in China?

5. At Math Tourney Prep School, there are 150 elementary and middle school students. If 40% of students are boys, $\frac{2}{3}$ of students are in middle school, and 30% of middle school students are boys, how many students are elementary school girls?

6. Since the number four is an unlucky number in China, buildings don't use the number four when they are numbering floors. For example, they don't have a floor labeled as "14" because 14 has a four in it. If the top floor of a building in China is labeled "50," how many floors does the building actually have?

7. Madilyn is playing a basketball game with her friends. In their basketball game, there are no free throws, so Madilyn can only make two-point shots or three-pointers. If she makes 12 shots, and scores 28 points, how many three-pointers did she make?

8. Matt and Linna were hungry one day, so they decided to have a chip eating-contest. Since Matt decides to be kind, he gives Linna a 60-second head start. Linna eats at a rate of 6 chips per minute, and Matt eats at a rate of 8 chips per minute. How many chips will Matt have eaten when he catches up to Linna?

3rd-4th Grade 2017 Answer Key

Sprint Round

1. 30 cents
2. $31
3. 14 inches
4. 19 eggs
5. 5 months
6. $2.50
7. $\frac{1}{6}$
8. 12 Lego pieces
9. 6 votes
10. 230 minutes
11. 1 inch
12. 6 lines of symmetry
13. 5 skirts
14. 16 tiles
15. 12
16. $\frac{1}{5}$
17. 36
18. Sunday
19. 14 desks
20. 0
21. 72 in^2
22. 26 cards
23. 55
24. 40

25. 50 marbles
26. 12 weeks
27. 18 ploops
28. 84 in^2
29. 32 jobs
30. 120 seconds

Target Round

1. 7 books
2. 10 triangles
3. 9 cups
4. 1,400,000,000 or 1.4 billion
5. 20 students
6. 36 floors
7. 4 three-pointers
8. 12 chips

5th-6th Grade 2014
Sprint Round

---

1. A certain poodle usually drinks water at a rate of 1 liter per 7 minutes. However, since the poodle is very thirsty today, its drinking rate is tripled. If we leave six-sevenths of a liter out for the poodle, in how many minutes will it finish?

2. Olivia likes to play basketball every other day of the week. If she starts playing on a Tuesday, how many times will she play basketball in two weeks, starting Tuesday?

3. If x is the greatest prime factor of 36 and y is the greatest prime factor of 27 then what is $x + y$?

4. Two concentric circles are drawn such that one circle's radius is 3 times the other. If the area between the two circles is 72π, what is the radius of the larger circle? (Two circles are concentric if they share the same center.)

5. When a turtle wants to cross a river, he has to pay $4. If he crosses the river 5 times on Monday, how much did he pay on Monday?

6. Serge has a crush on Pam and decides to buy her 5 yards of yarn (and some roses). However, the yarn can only be bought in inches. How many inches of yarn does he need to buy?

7. If a bottle of sunscreen costs $8 and Sue has a coupon that gives a 25% discount, how many dollars do she pay if she uses the coupon?

8. What is the smallest positive integer with four factors?

9. Allison took a series of tests during her freshman year. Her test scores were 92, 85, 87 and 100. However, instead of 100, her test should have been a 98.5. What is the new median? Express your answer as a decimal rounded to the nearest tenth.

10. I am a number between 1 and 100. You get the same number if you multiply me by 4 or add 198 to me. What number am I?

11. The width of a rectangular prism is 3. Its length and height are both 5. What is its surface area?

12. A dog runs toward a ball 30 feet away. If the dog is halfway there, how much farther does he have to run to get the ball and get back to his starting point (in feet)?

13. If the beaver slaps his tail on water 4 times in 5 seconds, then pauses for 5 seconds, then repeats the cycle, how many times does he slap his tail on the water in one minute?

14. If the lengths of two of the sides of an isosceles triangle are 6 and 12, then what is the perimeter of this triangle?

15. If there are 300 cubbyholes and Joe has 301 flyers, what is the probability that one cubby will have at least two flyers?

16. The length and width of a rectangle are both doubled. What is the percent increase between the new area and the original area?

17. A spinner has 24 sectors, numbered 1, 2, 3, ... 24. What is the probability that, after spinning the spinner once, it lands on either a prime number or an odd number? Express your answer as a common fraction.

18. In a total of 15 numbers, the median is 38. The largest number is 93. If the largest number is changed to 97, what is the median?

19. The product of all 4 sides of a square is 1296. What is the sum of all four of the square's sides?

20. If the cost of an apple and 2 oranges is 60 cents, the cost of an orange and 2 peaches is 36, the cost of a peach and 2 apricots is 120, and the cost of an apricot and 2 apples is 96, what is the cost of buying one of each (in cents)?

21. Three animals are arguing about a missing cookie. Frog says that Rabbit ate it, Rabbit says that Frog ate it, and Bird says that Frog is innocent. If exactly 2 animals are lying, who is sure to be innocent?

22. A man walks on a path around his house. Starting from his house, he first walks 30 feet east, then 40 feet north, then 60 feet west, then finally 80 feet south. How far, in feet, is he from his house?

23. John has some cookies. He gives half of them to his siblings, one-third of them to his parents, and one-twelfth to his friend. He eats the only cookie left. How many cookies did he start with?

24. If Josh took the square root of a positive number, tripled that number, and then added 8, he got the number 20. But instead of taking the square root of the number, he should have squared it. If Josh had squared the number, what would his result be?

25. Let $a \wr b = \frac{a|b-a|}{b}$, and $a \ddagger b = a \wr b + b \wr a$. What is $(4 \ddagger 6) \wr (6 \ddagger 4)$?

26. In the month of May there are 31 days. If May 4th is a Sunday, how many Sundays are there in May?

27. An 8-sided die is rolled (sides numbered 1-8), and then a 6-sided die is rolled. What is the probability that at least 1 prime number is rolled? Express your answer as a common fraction.

28. Ian is trying to draw a right triangle with a hypotenuse length of 15 and a leg length of 4. How long will the other leg be? Express your answer in simplest radical form.

29. The sum of 10 consecutive integers, starting with 11, equals the sum of 5 consecutive integers, starting with what number?

30. Annie is playing games of solitaire and listening to music at the same time. Each game of solitaire lasts for 5 minutes. She only wins a game if she hears her special song for some portion of that game. If her special song will play 30 minutes after she begins playing solitaire, Annie will win for the first time when she plays her nth game. What is the value of n?

Target Round

1. Sarah is delivering newspapers on a street of houses numbered 1 to 100 with even-numbered houses on the right and odd-numbered houses on the left. If she only delivers to house numbers that are a multiple of 3 and on the right, how many houses does she deliver newspapers to?

2. A room is one-half full of people. After 20 people leave, the room is one-third full. How many people would be in the room if it were full?

3. A cell phone battery is at 40% and will run out in 36 minutes. How long (in minutes) will the battery last if it is at 90%?

4. Paul is not very good at talking to girls. Every time Paul tries to talk to a girl, he gets very nervous and starts counting in multiples of 3. If Paul counts in multiples of 3 (3, 6, 9, 12, 15,...) until he reaches 1,242, how many numbers has Paul said?

5. In the month of April there are 30 days. How many possible ratios are there of the number of Mondays to the number of Thursdays in April? (April can start on any day of the week.)

6. Miki's age is 9 less than 2 times her sister's age. In 4 years, the sum of their ages will be twice their brother's current age, which is 5 more than her sister's current age. What is my age?

7. Robert, Amber, Charlie, Jasmine, Annie and Patricia were sitting next to each other (in that order) and one of them is holding a movie. The people at the ends are not holding the movie. The people sitting 2 spaces from either end are not holding the movie. The person holding the movie is no more than 3 spaces away from Patricia. Who is holding the movie?

8. Sally and her boyfriend Jason are going on a romantic picnic. Since Sally is very forgetful, he does not know what kinds of sandwiches Jason enjoys. To accommodate for this, she packs 3 types of breads, 5 types of cheeses (including some stinky ones!), and 6 types of meat. However, Jason refuses to have blue cheese with wheat bread, or white bread with roasted ham. How many choices of sandwiches does Jason have, if he must choose one bread, one cheese, and one meat?

5th-6th Grade 2014 Answer Key

Sprint Round

12. 45 feet
13. 24 times
14. 30
15. 100% or 1
16. 300%
17. $\frac{13}{24}$
18. 38
19. 24
20. 104 cents
21. Rabbit
22. 50 feet
23. 12 cookies
24. 776

25. 0

26. 4 Sundays

27. $\frac{3}{4}$

28. $\sqrt{209}$

29. 29

30. 7th game

Target Round

1. 16 houses

2. 120 people

3. 81 minutes

4. 414 numbers

5. 3 possible ratios

6. 13 years old

7. Annie

8. 79 sandwiches

5th-6th Grade 2015
Sprint Round

---

1. Susie has a rectangle. If she increases the length by 10% and width by 5%, by what percent is the area increased? Express your answer as a decimal rounded to the nearest tenth.

2. Bill has 5 unique Rubik's Cubes to give away to 4 people: Leon, Suhn, Mitchell, and Bill 2.0. How many ways can he distribute them?

3. What is the positive difference between $1+2+3+4+5+\ldots+100$ and $2+3+4+5+6+\ldots+101$?

4. A sequence is defined by $a_n = a_{n-1} \times n$. If a_5 is equal to 180, find the value of a_2.

5. The sides of a right triangle have lengths 8 cm, x cm, and 15 cm, and x is an integer. What is the area of the triangle?

6. In triangle ABC, angle A is equal to 65 and angle B is equal to 75. The bisector of angle C is drawn to meet AB at point D. What is the measure of angle ADC? (An angle bisector is a line that cuts an angle into two angles of equal measure.)

7. Five friends, Natalie, Lucy, Amy, Thomas, and Ian, go to Disneyland together. They each buy their own ticket for $156 each. They equally split up the lunch bill of $100 and they each buy one pair of Mickey Mouse ears for $22 a pair. If this is all they spent, how many dollars did each of them spend at Disneyland?

8. William, Zack, Bryan, Jason, and Joey are running in a race. These statements are true about the results:

   William finished before Jason but after Joey.

   Jason finished in third place.

   Bryan finished after Joey but he did not finish last.

   Zack finished after both Bryan and Jason.

   Who was the first place runner?

9. If the lengths of two sides of a isosceles triangle are 5 and 8 respectively, what is the sum of all the possible perimeters for the triangle?

10. Bob sells gum in packs of 3 and 5. What is the largest amount of gum that a person is not able to buy using a combination of Bob's packs of gum? For example, a person can get 14 pieces of gum by ordering 3 packs of 3 and 1 pack of 5.

11. Susan is a criminal case solver. Today, her case is to find out who ate which type of candy, Skittles, M&M's, Jolly Ranchers, and Twix. She knows Michael hates chocolate. Mindy says she would never eat Twix in a million years. John said that he saw a boy eat Skittles. Lastly, Joey is under a lot of pressure, and confesses that he ate the Jolly Ranchers. Given that nobody ate the same candy, what did Mindy eat? (Assume M&M's and Twix are the only two chocolate candies.)

12. Leon is playing Tic-Tac-Toe with Marcia Dai. Each game takes 16 minutes and 40 seconds. Assuming they don't take breaks how many games will they play in ten days?

13. Using a standard deck of cards, a class is doing a fun game. When a certain suit is a pulled from the deck, the class has to do a certain exercise. Hearts are jumping jacks, diamonds are sit-ups, clubs are push-ups, and spades are burpies. If all face cards (including aces) represent 10, and the number on the card represent how many of each exercise you need to do, how many sit-ups and push-ups would a student have completed after going through the whole deck?

14. If $x = 3y$ and $z = \frac{x}{2}$, find the value of $\frac{x+y+z}{x+y-z}$

15. With the letters I N D I A N A, how many ways are there to order the letters?

16. If 2 cans of paint are enough to paint a 1 yard x 1 yard square, how many cans are needed to paint the outside walls of a shed that is 10 yards long, 8 yards tall, and 4 yards wide? (The roof and ground do not need painting)

17. The Apple County Math Circle is choosing a committee of a president, vice president, and a secretary from a group of 30 members. How many committees can be made?

18. A store offers a free bucket of paint for every 3 buckets bought. Robert buys 10 buckets of paint, and Bob buys 6 buckets of paint. If each bucket of paint costs \$10 each, how much money is saved if they bought the paint together rather than separately?

19. In a party with 10 people, each person shakes hands once with everybody else. How many total handshakes occur?

20. The perimeter of an equilateral triangle is equal to that of a rectangle. If the side length of the triangle is 12, what is the maximum area the rectangle can have?

21. If $x + \frac{1}{x} = 10$, calculate $x^2 + \frac{1}{x^2}$.

22. Find $x+y$ if both x and y are whole numbers and $(2x+y)(x-2y)=7$ is satisfied.

23. They are having student elections at my school. There are 4 candidates for presidents, 3 candidates for vice president, and 5 candidates for activities commissioner. There were 3 candidates for historian, but one of them decided to opt out of the election, and two more joined in. How many different student bodies can there be?

24. What is the obtuse angle created by the hour hand and minute on an analog clock when it is 3:40?

25. If Matthew Tang destroys a card deck in 5 days while Arthur Liu destroys 3 decks in a week. How many days will it take for them to destroy 22 decks if they work together?

26. All terms in the following geometric sequence are positive integers. The first three terms are n, n + 3, and 2n + 6. What is the fourth term of this sequence?

27. We are playing chess. I am really bad, so I only have my bishop left, and it can only move on the black squares. The chess board that we are playing with is 6×6. What fraction of the board can the bishop not move on?

28. What is the maximum number of regions that 7 chords can make in a circle?

29. Let $ABCDEF$ be a 6-digit number, and let A, B, C, D, E, F be digits of different values. If $ABCDEF * 4 = EFABCD$, what is $ABCDEF$? (Assume A and E cannot equal 0.)

30. The length of one leg of a right triangle is 2 inches less than the length of the other leg. If the length of the hypotenuse is 10 inches, what is the length of the longer leg?

Target Round

---

1. What is the sum of the first 100 even natural numbers?

2. The cheese pizza at Cheesecake Factory was $12. For every topping I add, it costs an extra $1.20. I added pineapple, onions, and pine nuts. After eating my pizza, I gave the waiter 5% tip. If I used a gift card for $10, how much was my final bill (in dollars)? Express your answer as a decimal rounded to the nearest hundredth.

3. A strange clock chimes 1 time at 1 o' clock, 2 times at 2 o' clock, 3 times at 3 o' clock, 4 times at 4 o' clock, 5 times at 5 o' clock, and so on. How many times will the clock chime from 1-12 o' clock?

4. Using the numbers 1, 2, 3, 5, 8, how many three-digit, even numbers can be formed? (Digits cannot be repeated.)

5. Given that x, y, and z are positive numbers such that $x^2/2 = 4y$ and $8y = z^8$, what is x in terms of z?

6. If the number of seconds in 6 weeks can be expressed as $x!$, find x.

7. Find the area of a triangle with vertices at (2, 2), (3, 5), and (-1, 3).

8. The lengths of three sides of a triangle are 1027, 2014, and x. If the possible values of x are expressed as $m < x < n$, what does $m + n$ equal?

5th-6th Grade 2015 Answer Key

Sprint Round

1. 15.5%
2. 1024 ways
3. 100
4. 3
5. 60
6. 95°
7. $198
8. Joey
9. 39
10. 7 pieces of gum
11. M&Ms
12. 864 games
13. 188
14. $\frac{11}{5}$
15. 630 arrangements
16. 448 cans
17. 23,460 possible combinations
18. $10
19. 45 handshakes
20. 81
21. 98
22. 4
23. 240 student bodies
24. 130°

25. 35 days
26. 24
27. $\frac{1}{2}$
28. 29
29. 142,857
30. 8 inches

Target Round

1. 10,100
2. $6.38
3. 78
4. 24
5. z^4
6. 10
7. 5
8. 4028

5th-6th Grade 2016

Sprint Round

---

1. You start with 20 apples. You eat half of your apples. Your friend gives you 6 more apples. How many apples do you have?

2. A parallelogram has an area of 15 cm^2 and height of 7.5 cm. What is the length of its base?

3. $3x - 4y = 18$ and $4x - 3y = 10$. Compute $x - y$.

4. A square and an equilateral triangle have side length 5. Michelle glues one edge of the equilateral triangle to one edge of the square, keeping the square and the triangle touching at only one edge. What is the perimeter of this new figure?

5. How many lines of symmetry does the following figure have?

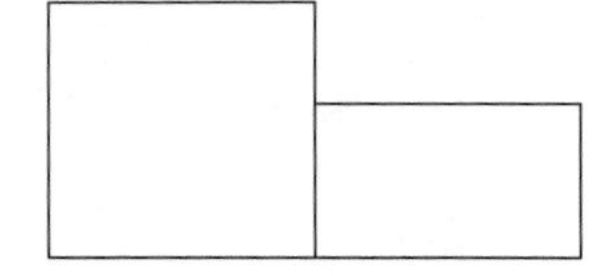

6. The angles of one triangle are $2x^\circ$, $3y^\circ$, and 80°. The angles of another triangle are 120°, x°, and y°. What is $x + y$?

7. Clarisse flips a coin twice. What is the probability that the two flips yield different results?

8. Suppose that $a \diamond b = \frac{ab}{a+b}$. Compute $2 \diamond (3 \diamond 6)$.

9. A school has some books. All the books can be split evenly among 6 classrooms and among 10 classrooms. What is the smallest number of books that the school can have?

10. Allison has to solve 15 math problems for homework. 5 are easy problems, 5 are medium problems, and 5 are hard problems. If it takes her 2 minutes to solve an easy problem, 4 minutes to solve a medium problem, and 7 minutes to solve a hard problem, how many minutes does it take her to finish her homework?

11. Two chipmunks can build a treehouse in five hours. How much time will it take for ten chipmunks to build twenty treehouses?

12. In rectangle $ABCD$, AB is 1 less than twice the length of BC. If the perimeter is 22, what is the area of $ABCD$?

13. Matt is sleeping through his math final, which happens to be 100 questions long. When he is woken up with 15 minutes remaining, he starts, but cannot finish his test. If Matt had answered three-fifths of the questions on his math final, how much longer would he have needed? Express your answer in minutes.

14. The sum of the area and the perimeter of a rectangle with integer sides is 38. What is the largest possible area of the rectangle?

15. Michael and Jason have a pizza eating competition. The pizzas are perfectly shaped squares with side lengths that are whole numbers greater than 1. If the total area of one pizza Michael ate is 25 in^2 less than the total area of one pizza Jason ate, what is the side length of the larger pizza?

16. Michael was very salty about losing the first eating competition against Jason, and called for a rematch. This time, they are eating extremely salty square saltine crackers. Half of the area of each salty square saltine cracker is covered in pure salt. If Michael eats 24 in^2 more salty saltine crackers than Michael, how much more salt does he consume?

17. Jason is a math student who has trouble reading numbers- he flips the tens and ones digits of every number he sees. For example, he reads 24 as 42, and 63 as 36. If he calculated the area of a 51 by 12 rectangle, what is the positive difference between his answer and the correct area?

18. There are two rectangles with integer side lengths, both with area 48. What is the greatest possible difference between the perimeters of the two rectangles?

19. Anh is trying to decide which books to read next, and so goes to her local bookstore. The bookstore has three sections: sci-fi, fantasy, and non-fiction, each with 10 books available. If she wants to choose two books from two different sections, how many ways can she choose them?

20. A, B, C, D, E, and F are digits from 1 to 9 and ABC and DEF represent 3 digit numbers. If $ABC + DEF = 1024$, what is the maximum possible value of $CBA + FED$?

21. Twelve girls are trying to figure out how to split up into two teams of six for the All Girls Math Tournament. 3 of them are from Generic Middle School, 4 are from Standard Middle School, 2 are from Normal Middle School, and the other 3 are homeschooled. If each non-homeschooled girl must be on the same team as the other girls from her school, how many ways are there to split them up into two teams?

22. Aleena has a collection of 12 identical lavender-colored Sharpie markers, and she wants to distribute them all among three of her friends. If each friend must receive at least 3 Sharpie markers, how many ways are there to distribute the markers?

23. The first two terms of an arithmetic series are 5 and 10. The first two terms of a geometric series are also 5 and 10. What is the positive difference between the 5th terms of each series?

24. Points A, B, C, and D are on a line, from left to right in that order. B is the midpoint of AC, and C is the midpoint of AD. If point E is not on the same line, what is the ratio of the area of $\triangle ABE$ to the area of $\triangle ADE$? Express your answer as a common fraction.

25. The library is a very popular place in the small town of Bookworm. Every other day, Anika goes there work on her homework. Every third day, Ben goes there to read. Every fifth day, Cassandra goes there to meet up with the local book club. Given that all three were there on December 31 last year, how many days were *exactly* two of them there in January this year?

26. Triangle ABC with non-zero area has $AB = 42$ and $BC = 73$. If AC has integer length, how many different possible values can it be?

27. The coordinates of a polygon are (0,0); (1,4); (5,0); (2,4); (2,2); and (5,2). What is the area of the polygon?

28. Betty tosses a fair coin 6 times. What is the probability that she gets strictly more heads than tails?

29. Right triangle ABC has $AC = 8$ and $CB = 6$. M is the midpoint of AB. Pick point N on line CM with M between C and N such that $\angle CAB = \angle BAN$. Compute MN. Express your answer as a common fraction.

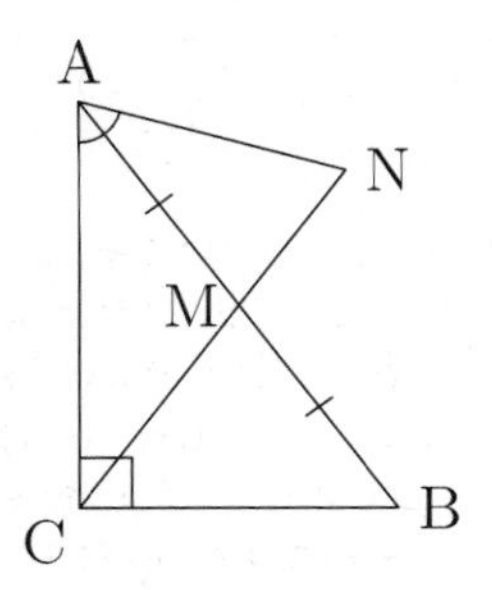

30. Two circles, O_1 and O_2, of radius 6 and 8, respectively, intersect at points P and Q, as shown. AB is a line segment that passes through P and with one end on each circle. AQ is tangent to $\odot O_2$, and BQ is tangent to $\odot O_1$. Find the area of $\triangle AQB$.

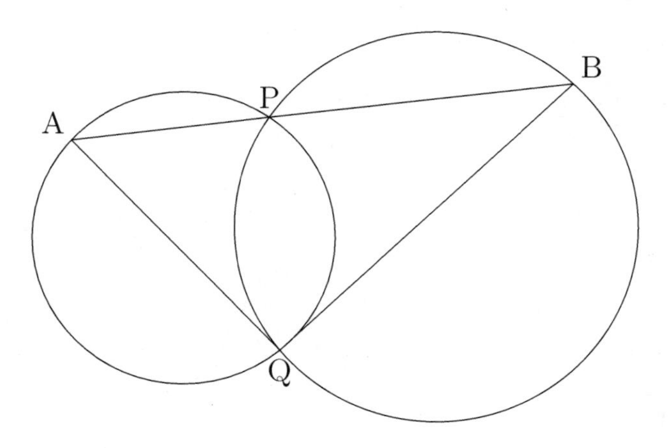

Target Round

1. The area of a square is 12.25 in^2. What is the perimeter?

2. Bessie has 7 friends who are either cows or chickens. In total, her friends have 22 legs. How many of Bessie's friends are cows?

3. Adam uses sticky notes to remind him to do his homework. He uses 1 sticky note on the first day of school, 2 sticky notes on the second day of school, and so on, using twice as many sticky notes each day. How many sticky notes does he use in total by the end of the sixth day of school?

4. Tom and Laura have a total of 17 apples. The number of apples that Tom has is divisible by 4. The number of apples that Laura has is divisible by 3. How many apples does Laura have?

5. The perimeter of a regular pentagon is equivalent to the perimeter of an equilateral triangle. If the side length of the pentagon is 6, what is the side length of the triangle?

6. A "holey number" is defined as a whole number whose digits consist only of 0, 4, 6, 8, and 9 (digits may be repeated). How many 3-digit holey numbers are there?

7. Anne the Ant is walking on a line. Each second, she flips a coin. If it lands heads up, she walks a step forwards. Otherwise, she stays where she is. What is the probability that, after 6 seconds, she is exactly 3 steps away from where she started? Express your answer as a common fraction.

8. A six-sided die is weighted so that each odd number is equally likely, and each even number is equally likely as well. If the expected value of one roll is $3\frac{2}{3}$, what is the probability of rolling a one?

5th-6th Grade 2016 Answer Key

Sprint Round

1. 16 apples
2. 2 cm
3. 4
4. 25
5. 0 lines of symmetry
6. $40°$
7. $\frac{1}{2}$
8. 1
9. 30 books
10. 65 minutes
11. 20 hours
12. 28
13. 10 minutes
14. 20
15. 13 inches
16. 12 in^2
17. 297
18. 70
19. 300 combinations
20. 1519
21. 4 ways
22. 10 ways
23. 55
24. $\frac{1}{4}$

25. 7 days

26. 83 side lengths

27. 12

28. $\frac{11}{32}$

29. $\frac{125}{39}$

30. 96

Target Round

1. 14

2. 4 cows

3. 63 sticky notes

4. 9 apples

5. 10

6. 100 "holey numbers"

7. $\frac{5}{16}$

8. $\frac{1}{9}$

5th-6th Grade 2017

Sprint Round

---

1. Susie went to the movies and brought twenty dollars. Her friends doubled the amount of money she had. Then she spent \$6 on popcorn and \$3 on a soda. How much money did Susie have left?

2. Jenna was making a birthday cake, so she bought 2 dozen eggs. If she had 5 eggs left over, how many eggs did she use to make the cake?

3. To participate for the All Girls Math Competition, Casey needs to buy some pencils and erasers. A pencil costs 49 cents and an eraser costs 99 cents. If Casey wants to buy 2 pencils and 2 erasers, how much money would Casey need (in cents)?

4. Prom King and Queen are being decided at the All Girls Math Tournament Annual Dance. Jane received 5 more votes than Karen's amount of votes tripled. If Jane received 23 votes, then how many votes did Karen receive?

5. At the end of school, Ms. Johnson decides to hand out 30 cookies among her class. There are 25 students in her class. If 20% of her class is allergic to cookies, how many cookies should each student get? (Answer with a fraction in lowest terms.)

6. The apps Temple Run and Candy Crush take up a total of 360 MB of space on Alice's phone. If Temple Run is three times as big as Candy Crush, how large is Candy Crush? (Note: A MB, or megabyte, is a unit that measures storage on electronic devices.)

7. In a magic square, the sum of the numbers in each vertical, horizontal, and diagonal row must add up to the same number. Also, each number from 1 to 9 (including 1 and 9) must be used, and they can only be used once. In the magic square below, what number belongs in the highlighted box?

2		4
(highlighted)	5	
		8

8. One day, Ms. Chou decided to measure the height of her eleven students. If their heights are 52 in, 53 in, 49 in, 56 in, 52 in, 54 in, 55 in, 53 in, 57 in, 47 in, and 52 in, what is the difference between the median height and the mode height, in inches?

9. In a regular deck of 52 cards, what is the probability of drawing an Ace of any suit? (Answer with a fraction in lowest terms.)

10. There are about 7 billion people in the world, and about 60% of them live in Asia. If $\frac{2}{3}$ of Asians do **NOT** live in China, about how many people live in China?

11. Billy Jean draws 5 squares side by side, as shown in the figure below. If the squares have side lengths of 1, 2, 3, 4, and 5, in that order, what is the area of the figure?

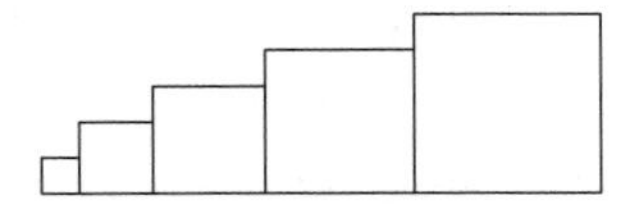

12. What is the perimeter of Billy Jean's figure in the previous problem?

13. Matt and Linna were hungry one day, so they decided to have a chip eating-contest. Since Matt decides to be kind, he gives Linna a 30-second head start. Linna eats chips at a rate of 90 cm^2 per minute, and Matt eats at a rate of 8 chips per minute. If each chip is a triangle with a 5 cm base and 6 cm height, how many chips will Matt have eaten when he catches up to Linna?

14. Linna was mad about losing the eating contest to Matt, so she decided to eat some cupcakes to make herself feel better. Each cupcake is 60 cm^3, and a quarter of the cupcake is frosting. If Linna ate 75 cm^3 of frosting, how many cupcakes did she eat?

15. When Amy was hunting for Easter eggs, she found 30 eggs, each with 2 pieces of candy inside. She was really excited, so she immediately ate $\frac{1}{4}$ of her candy, and then hid the rest in her secret stash. However, her brother John found the secret stash and took an unknown amount of candy. Amy was sad that her brother stole her candy, so she ate $\frac{2}{3}$ of her remaining candy to cheer herself up. If there were 10 pieces of candy left afterwards, how many pieces of candy did her brother take?

16. Kevin has an obsession with pandas, so he decided to make a WALL OF PANDAS by purchasing lots of framed pictures of pandas. Each picture is 3 ft long and 2 ft tall. If Kevin's WALL OF PANDAS is 13 ft long and 11 ft tall, how many complete pictures can he possibly fit on his wall without rotating any of the pictures?

17. Rosie is buying a new TV from TV Emporium, so she brought two coupons. The first coupon would give her a 20% discount, while the other coupon would give her a 10% discount and an additional $40 off. If each coupon would save her the same amount of money, how much would her new TV cost without using any coupons?

18. Casey draws five points on the same piece of paper. Then, she draws lines connecting each point to every other point. When she does this, how many lines would she have drawn? (Assume that no three points are on the same line.)

19. If ten people working for ten hours can complete ten jobs in total, then how many jobs can twenty people complete in twenty hours?

20. A circle is drawn inside a square so that the circle is touching the center of each side of the square. If the side length of the square is 6, what is the area of the circle? (Please leave answer in terms of π.)

21. On Maddie's first four tests, she received a 96, an 88, an 84, and a 93. If Maddie is about to take a fifth test, and wants to get an average of 91, what score does she have to get on the final test?

22. When a bouncy ball drops from a certain height, it will bounce back up to only half of that height. If that ball is dropped from 1 meter high, after which bounce will the ball bounce to a height that is less than a centimeter?

23. Sophia draws a house with a square body, a triangular roof, and a chimney in the shape of a trapezoid. Using the picture below, find the area of the house.

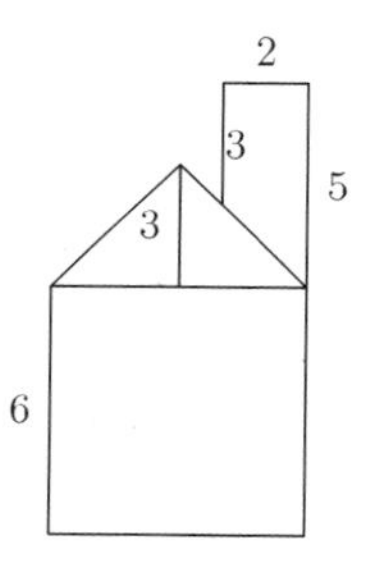

24. At the All Girls Car Wash, Anne and Bella are both washing cars. If Anne can wash a car in 30 minutes by herself and Bella can wash a car in 20 minutes by herself, how many minutes will it take them to wash a car together?

25. Andrea, Bethany, Caroline, and Daisy went to the movie theater together, and decided to sit in four seats in a row. If Andrea and Bethany are mad at each other and CANNOT sit next to one another, how many different ways can the four girls arrange themselves?

26. Fifi was thirsty one day, so she bought three cartons of milk. Before she went to bed, she put the milk cartons in fridge so they wouldn't spoil. However, the fridge was broken, so each carton of milk has a

$\frac{1}{3}$ chance of spoiling overnight. What is the probability that *none* of the cartons will spoil overnight?

27. At 11:00 AM, Train Amnesty left the station, going north at a speed of 250 kilometers per hour. At 12:00 PM, the bullet train left the station, going north at a speed of 350 kilometers per hour. At what time will the bullet train catch up to Train Amnesty?

28. Drew and Jenna have a deck of six cards numbered 1 through 6. Drew shuffles the deck, picks the top card, puts his card back in, and shuffles the deck. Jenna then picks a card, and they multiply their two numbers together. If the result is odd, Drew wins the game, and if the result is even, Jenna wins the game. What is the probability Jenna wins the game? (Answer with a fraction in lowest terms.)

29. Right now, Alexa is four times as old as her brother Michael. In four years, she'll be twice as old as Michael. How old will Alexa be 10 years from now?

30. A mouse has a block of cheese that is in the shape of a triangular prism. The base of the block of cheese is in the shape of a right triangle with legs that are 6 inches long and 8 inches long. The height of the block is 4 inches. If the mouse ate $\frac{1}{3}$ of the cheese, how much cheese is left (in cubic inches)?

Target Round

1. Jessica just turned 16, so she can finally buy her first car! She went to the car dealership and found two cars she really liked, both of which were on sale. The pink car was \$35,000, but the dealer took \$3,000 off the price, and then took 20% off the discounted price. The purple car was \$40,000, but the dealer made it 25% off of the original price, then discounted it by another \$2,000. What is the difference in price between the two cars?

2. Find the area of the shaded region. (Leave your answer in terms of π.)

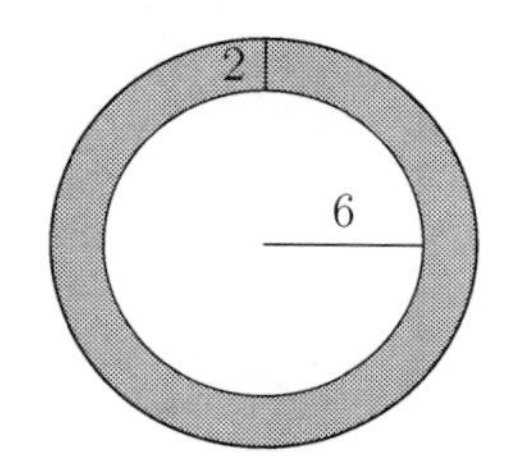

3. What is the area of the shaded region?

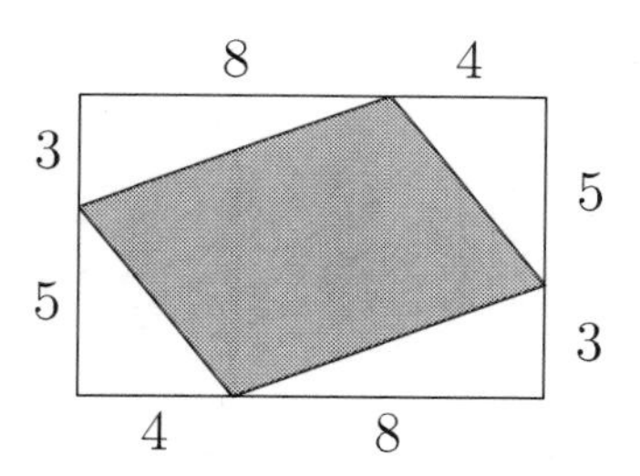

4. Candace and Stacy are playing a game. Candace picks a card from a deck of six cards numbered 1 through 6. Stacy picks a card from a deck of four cards numbered 1 through 4. Whoever has the higher number wins the game. What is the probability that Stacy will pick the higher number? (Answer with a fraction in lowest terms.)

5. Joey was at the bookstore and wanted to buy some Harry Potter novels and some Percy Jackson novels. If she bought one book from each series, it would cost her \$35. If she bought all seven Harry Potter books and all five Percy Jackson books, it would cost her \$215. How much does one Harry Potter book cost? (Assume each book in a series is the same price.)

6. What is the number parallelograms in the figure below?

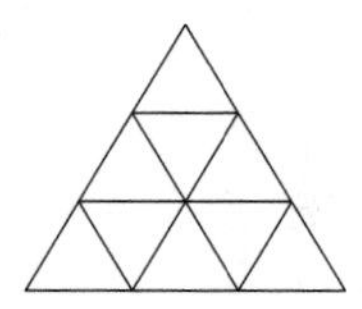

7. Elena has four numbered doors in front of her, each with a corresponding guard in front of it. She knows that one of the doors has treasure behind it, but she's does not know which door it is behind, so she decides to ask the guards some questions. However, EXACTLY one of these guards is lying. This is how they replied:

   Guard 1: The treasure is behind the door of the guard who is lying.

   Guard 2: Guard 3 is a liar.

   Guard 3: The treasure is behind Door #1.

   Guard 4: I am not a liar.

   Given their responses, which door is the treasure behind?

8. Michelle went outside to get her mail, but she realized that the mailman forgot to deliver her mail! Luckily, she sees that the mail truck is 25 meters down the road, so she starts walking toward it at a speed of 2 m/s. However, the mail truck doesn't see her, so it drives away from her speed of 5 m/s, but has to take 10 seconds to stop at a house and deliver its mail. If each house's mailbox is 25 meters away from each other, and the mailman just finished delivering someone's mail, how long will it take for Michelle to catch up to the mail truck?

5th-6th Grade 2017 Answer Key

Sprint Round

1. $31
2. 19 eggs
3. 296 cents
4. 6 votes
5. $\frac{3}{2}$ cookies
6. ~~270~~ 90 MB
7. 7
8. 1 inch
9. $\frac{1}{13}$
10. 1,400,000,000 or 1.4 billion
11. 55
12. 40
13. 12 chips
14. 5 cupcakes
15. 15 pieces
16. 20 pictures
17. $400
18. 10 lines
19. 40 jobs
20. 9π
21. 94%
22. 7 times
23. 53
24. 12 minutes

25. 12 ways
26. $\frac{8}{27}$
27. 2:30 PM
28. $\frac{3}{4}$
29. 18 years old
30. 64 in^3

Target Round

1. $2,400
2. 28π
3. 52
4. $\frac{1}{4}$
5. $20
6. 15 parallelograms
7. Door #3
8. 75 seconds

Solutions

3rd-4th Grade 2014 Solutions
Sprint Round

---

1. How much does Alice pay if she buys 42 computers for \$700 each?

   Solution: If Alice buys 42 computers at 700 dollars each, then the total cost of the computers is $42 \times 700 = 42 \times 7 \times 100 = 29,400$. Therefore, Alice pays $\boxed{\$29{,}400}$ for 42 computers.

2. What is $\frac{2}{9} \div \frac{8}{3}$? Express your answer as a common fraction.

   Solution: In order to solve this problem, realize that division is the inverse of multiplication. This means that multiplying the first fraction by the reciprocal of the second fraction is the same as dividing the first fraction by the second fraction.

   Applying this logic to this problem, dividing $\frac{2}{9}$ by $\frac{8}{3}$ is the same as multiplying $\frac{2}{9}$ by the reciprocal of $\frac{8}{3}$. The reciprocal of a number is 1 divided by the number. For example $1 \div \frac{8}{3}$, or $\frac{3}{8}$, is the reciprocal of $\frac{8}{3}$. When finding the reciprocal, switch the numerator and denominator.

   Next, multiply the first number by the reciprocal of the second. To multiply $\frac{2}{9}$ by $\frac{3}{8}$, multiply the numerator from the first fraction by the numerator from the second fraction and multiply the denominator from the first fraction by the denominator from the second fraction. This means that $\frac{2}{9} \times \frac{3}{8} = \frac{6}{72} = \boxed{\frac{1}{12}}$

3. A baby is crying at a rate of 2 tears per second. How many tears come out after 3 minutes of crying?

   Solution: First, find how many tears the baby cries in a minute. Since it cries 2 tears every second, and there are 60 seconds in a minute, it will cry $2 \times 60 = 120$ tears in a minute.

   Then, to find out how many tears are cried in 3 minutes, multiply 3 by the number of tears the baby cries in a minute, which would be $3 \times 120 = \boxed{\text{360 tears}}$.

4. John is buying cookies. If they cost 35 cents per cookie, how much (in dollars) do 11 cookies cost? Express your answer as a decimal rounded to the nearest hundredth.

   Solution: First, find the total cost in cents. If there are 11 cookies, and they each cost 35 cents, the total cost of the cookies is $11 \times 35 = 385$ cents.

Then, realize that 100 cents make a dollar, or that $3 \times 100 = 300$ cents makes 3 dollars. Next, by subtracting 3 dollars (or 300 cents) from 385 cents, there are $385 - 300 = 85$ cents left over. This means the cost of the cookies is \$3.00 + 85 cents, or $\boxed{\$3.85}$.

5. A wave of minions is eating bread. If they finish one loaf of bread in 8 seconds, how many minutes does it take for them to finish 45 loaves of bread?

   Solution: To start, find how many seconds it would take for the the minions to eat 45 loaves of bread. If the minions eat bread at a rate of 8 seconds per loaf, it would take $8 \times 45 = 360$ seconds for the minions to eat 45 loaves.

   However, this problems asks for how many *minutes* it takes the minions to eat the loaves of bread. Note that there are 60 seconds in a minute. To convert minutes to seconds, take the total amount of seconds, in this case 360, and divide it by 60. Therefore, it takes $360 \div 60 = \boxed{6 \text{ minutes}}$ for the minions to eat all of the bread.

6. What is $9 - 8 + 7 - 6 + 5 - 4 + 3 - 2 + 1$?

   Solution: For this problem, split the this equation down into smaller parts, like so:

$$\begin{aligned} &9 - 8 + 7 - 6 + 5 - 4 + 3 - 2 + 1 \\ =\ &(9 - 8) + (7 - 6) + (5 - 4) + (3 - 2) + 1 \\ =\ &1 + 1 + 1 + 1 + 1 = \boxed{5}. \end{aligned}$$

7. If 1400 cushions are as heavy as 3500 pillows, then how many cushions are as heavy as 5000 pillows?

   Solution: For this problem, first use a ratio to compare the weight of cushions to the weight of the pillows. The ratio of cushions to pillows is $\frac{1{,}400 \ cushions}{3{,}500 \ pillows}$, which simplifies to $\frac{2 \ cushions}{5 \ pillows}$. This means 2 cushions weigh the same as 5 pillows.

   Next, since 5 pillows weigh the same as 2 cushions, then $5 \times 1000 = 5000$ pillows are as heavy as $2 \times 1000 = 2000$ cushions. Therefore, $\boxed{2000 \text{ cushions}}$ are as heavy as 5000 pillows.

8. Jamie has a .554 batting average. Allison has a .336 batting average. What is Jamie's average minus Allison's average? Express your answer as a decimal rounded to the nearest thousandth.

   Solution: To solve this problem, write out the Jamie's average above Allison's average, lining up the decimals with each other. Then subtract normally as if you were subtracting 554 from 336. However, once you have subtracted the numbers, make sure to carry down the decimal point into the answer. As shown by the subtraction equation below, Jamie's average minus Allison's average is $\boxed{0.218}$.

$$\begin{array}{r} {\scriptstyle 4\;14} \\ 0.5\cancel{5}\,\cancel{4} \\ -0.3\,3\,6 \\ \hline 0.2\,1\,8 \end{array}$$

9. What is the remainder when $820 \div 17$?

   Solution: Use long division:

$$\begin{array}{r} 48 \\ 17\overline{)820} \\ \underline{68} \\ 140 \\ \underline{136} \\ 4 \end{array}$$

   Therefore, the remainder is $\boxed{4}$.

10. What time is it 3000 seconds after 1:30?

    Solution: First, convert the seconds to minutes. Since there are 60 seconds in a minute, divide the total seconds, in this case 3000, by 60. This means there are $3000 \div 60 = 50$ minutes in 3000 seconds.

    Next, add 50 minutes to 1:30. However, adding 50 minutes to 1:30 results in 1:80, which is impossible. In order to fix this, subtract 60 minutes from the minutes side, and add an hour to the hour side. This means that 1:80 is equal to $\boxed{2{:}20}$, which is the final answer.

11. Helium has 2 protons, Beryllium has 4 protons, Barium has 56 protons, and Mercury has 80 protons. If I have two Heliums, one Beryllium, one Barium, and four Mercuries, how many protons do I have?

    Solution: To start, find the amount of protons from each element. Since there are two Heliums with two protons each, that means that Helium makes up $2\times2 = 4$ of the protons. Next, there is one Beryllium with 4 protons, meaning that Beryllium is responsible for 4 protons. Next, one Barium means there are 56 protons form Barium. Lastly, there are 80 protons in one Mercury, and with four Mercuries there are $80 \times 4 = 320$ protons from Mercury.

    Finally, add up all the protons from each element to find the total number of protons: $4 + 4 + 56 + 320 = \boxed{384 \text{ protons}}$.

12. How many positive factors does the number 14 have?

    Solution: A factor is any number that can be evenly divided into another number. The positive factors of the number 14 are 1, 2 and 7, and 14 because these are the only numbers that can divide into 14 without leaving a remainder. This means 14 has a total of $\boxed{4 \text{ factors}}$.

13. Jamie is cooking 3 pounds of chicken. If he has to wait 35 minutes for every pound of chicken to cook, how many minutes does he have to wait for all of his chicken to finish cooking?

    Solution: To find how many minutes Jamie must wait for his chicken, multiply the amount of time needed to cook each pound, in this case 35 minutes, by the total amount of pounds, which is 3. Therefore, it will take $35 \times 3 = \boxed{\text{105 minutes}}$ for Jamie to cook 3 pounds of chicken.

14. Suppose John wants to cut his huge ruler, which is 144 inches long, into 12 different rulers, which are each 12 inches long. How many cuts does he make in his huge ruler?

    Solution: To make this problem easier, it helps to visualize the problem by drawing it on a piece of paper. If John makes one cut, he'll have two pieces. If he make two cuts, he'll have three pieces. If he makes three cuts, he will have four pieces, and so on. Continuing this pattern, if John makes $\boxed{\text{11 cuts}}$, he will have twelve pieces.

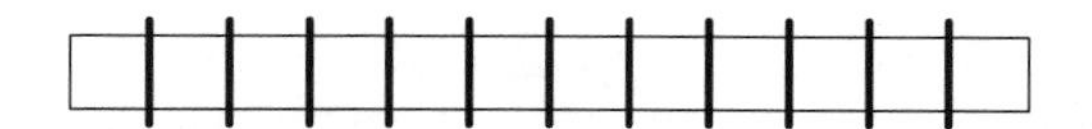

15. If an ice cube tray is made up of a 3×9 grid of squares. Each square can contain one ice cube. How many ice cubes can be made from 3 trays?

    Solution: First, find how many ice cubes each tray holds. If there are three rows of nine ice cubes each, then there are a total of $3 \times 9 = 27$ ice cubes per tray.

    Next, since there are three ice cube trays, the trays can hold $27 \times 3 = \boxed{\text{81 ice cubes}}$ altogether.

16. The Problem Writing Committee writes 10 problems in 1 minute. How many seconds does it take for the committee to write 16 problems?

    Solution: Let x be the amount of time it take to write 16 problems. If the committee writes 10 problems in one minute, then it writes 10 problems in 60 seconds. To find the amount of seconds it takes to write 16 problems use a proportion to solve:

$$\frac{60\ sec}{10\ problems} = \frac{6\ sec}{1\ problem} = \frac{x\ sec}{16\ problems}$$

    This proportion shows that it takes $6 \times 16 = \boxed{\text{96 seconds}}$ to write 16 questions.

17. Sarah's house has 5 stories. If the first story has 10 windows, and the following stories have one less window than the story below it, how many windows does Sarah's house have?

    Solution: The first story of Sarah's house has 10 windows. This means there must be 9 windows in the second story, 8 windows in the third story, 7 windows in the fourth story, and 6 windows in the fifth story. Adding the number of windows on each floor together, Sarah's house has a total of $10 + 9 + 8 + 7 + 6 = \boxed{\text{40 windows}}$.

18. Mary keeps all of her money in a piggy bank. On Monday, she takes out one dollar. On Tuesday, she puts in two dollars. On Wednesday, she removes four dollars. On Thursday, she puts in eight dollars. If she has fifty dollars in the bank on Friday, how many dollars did she have before Monday?

    Solution 1: To find out how much money Mary had in the bank before Monday, start with the money she had on Friday, and work backwards. If Mary had \$50 in the bank on Friday, and put \$8 in the bank on Thursday, there must have been $\$50 - \$8 = \$42$ in her bank on Wednesday night.

    On Wednesday, she takes out four dollars, so she had $\$42 + \$4 = \$46$ in her bank on Tuesday night.

    On Tuesday, she adds two dollars to her piggy bank, which means she had $\$46 - \$2 = \$44$ in her bank on Monday night.

    On Monday, she takes out one dollar, so she had $\$44 + \$1 = \boxed{\$45}$ in her bank before Monday.

    Solution 2: Let m be the original amount of money Mary has in her piggy bank.

    On Monday, she takes out one dollar, leaving $m - 1$ dollars left in her piggy bank.

    On Tuesday, she adds two dollars to her piggy bank, which gives her a total of $(m - 1) + 2 = m + 1$ dollars in the piggy bank.

    On Wednesday, she takes out four dollars, leaving her with a total of $(m + 1) - 4 = m - 3$ dollars in her bank.

    Finally, on Thursday she adds eight dollars to her piggy bank, which means she has $(m - 3) + 8 = m + 5$ dollars in her bank.

    Since she has 50 dollars in the bank on Friday, and no extra money was added after Thursday, $m + 5 = 50$. Next, subtract 5 from both sides of the equation to get that $m = 45$. This means that before Monday, Mary had $\boxed{\text{45 dollars}}$.

19. The mad scientist Eric is testing his new invention, the Best Enlarging Ray Gun, or BERG. The BERG enlarges objects by a factor of 8

in all dimensions. If Eric fires the BERG at a popsicle that is originally 3.5 inches long, what is the final length of the popsicle (in inches)?

Solution: To increase the length of the popsicle by a factor of 8, multiply the length by 8. This means the new length of the popsicle is $3.5 \times 8 = \frac{7}{2} \times 8 = 7 \times 4 = \boxed{28 \text{ inches}}$.

20. If x is the greatest prime factor of 36 and y is the greatest prime factor of 27 then what is $x + y$?

Solution: To start, find the greatest prime factor of 36. The greatest prime factor is the largest prime number that divides evenly into the given number. Now, find all the factors of 36. The factors of a number are numbers that when multiplied with another number produces the original number. The factors of 36 are 1, 2, 3, 4, 6, 9, 12, 18, and 36. Next, pick out the greatest prime number from this series. A prime number is a number whose only factors are one and itself. The greatest prime number in this series is 3. That means $x = 3$.

Next, find the greatest prime factor of 27. The factors of 27 are 1, 3, 9, 27. The greatest prime factor of 27 is 3 as well, meaning that $y = 3$.

Now, plug the numbers into the equation: $x + y = 3 + 3 = \boxed{6}$.

21. A certain poodle usually drinks water at a rate of 1 liter per 7 minutes. However, since the poodle is very thirsty today, its drinking rate is tripled. If we leave six-sevenths of a liter out for the poodle, in how many minutes will it finish?

Solution: The normal rate the poodle drinks water at is 1 liter every 7 minutes, or $\frac{1}{7}$ liters per minute. However, since the rate triples, it drinks at three times the normal rate, or $3 \times \frac{1}{7} = \frac{3}{7}$ liters per minute. Since the bowl has $\frac{6}{7}$ liters of water, and the poodle drinks $\frac{3}{7}$ liters every minute, it will take the poodle $\frac{6}{7}\ liters \div \frac{3}{7}\frac{liters}{min} = \boxed{2 \text{ minutes}}$ to drink all the water.

22. What is the value of $5\% + 0.05 + \frac{5}{100}$? Express your answer as a decimal rounded to the nearest hundredth.

Solution: In order to make this problem easier, convert the 5% into and $\frac{5}{100}$ into a decimal. A percent is one hundredth. In this case, 5% is equal to $\frac{5}{100}$, which is equal to 0.05. Now, add all the values together. This gives the value of $0.05 + 0.05 + 0.05 = \boxed{0.15}$.

23. Olivia likes to play basketball every other day of the week. If she starts playing on a Tuesday, how many times will she play basketball in two weeks, starting Tuesday?

Solution: Since there are seven days in one week, there are 14 days in two weeks. If Olivia only plays basketball every other day, she plays basketball half the time. If she plays basketball one-half of the time in 14 days, she plays basketball $\frac{1}{2} \times 14 = \boxed{7 \text{ times}}$ in two weeks.

24. When a turtle wants to cross a river, he has to pay $4. If he crosses the river 5 times on Monday, how much did he pay on Monday?

    Solution: If he crosses five times, and he has to pay $4 each time he crosses, he pays a total of $\$4 \times 5 = \boxed{\$20}$.

25. Serge has a crush on Pam and decides to buy her 5 yards of yarn (and some roses). However, the yarn can only be bought in inches. How many inches of yarn does he need to buy?

    Solution: First, find how many inches are in one yard. Since there are 3 feet in a yard, and 12 inches in one foot, there are $12 \times 3 = 36$ inches in a yard. This means if Serge needs to buy 5 yards of yarn, he is buying $36 \times 5 = \boxed{180 \text{ inches}}$ of yarn for Pam.

26. Allison took a series of tests during her freshman year. Her test scores were 92, 85, 87 and 100. However, instead of 100, her test should have been a 98.5. What is the new median? Express your answer as a decimal rounded to the nearest tenth.

    Solution: To calculate the median, list all the numbers in numerical order, and the number in the middle of the list is the median. However, if there are an even amount of numbers, the median is the average of the two middle numbers.

    First, start by listing her accurate tests scores in numerical order, like so: 85, 87, 92, 98.5. Since there are an even amount of numbers, the median is the average of the two middle numbers, which are 87 and 92. The average of 87 and 92 is $(87 + 92) \div 2 = \boxed{89.5}$.

27. A dog runs toward a ball 30 feet away. If the dog is halfway there, how much farther does he have to run to get the ball and get back to his starting point (in feet)?

    Solution: First, find how much further the dog has to run to make it to the ball. If the dog is halfway there, and the ball is 30 feet away, that means the dog has moved $30 \div 2 = 15$ feet. This also means that the dog still has 15 feet to go until he reaches the ball.

    Next, once the dog reaches the ball, he must return back to his original position. Since the ball was 30 feet away from his starting point, he has to run 30 feet to get back to where he started.

    Lastly, add the 15 meters it took to reach the ball to the 30 meters it took to get back to the starting point: $15 + 30 = \boxed{45 \text{ feet}}$.

28. If the beaver slaps his tail on water 4 times in 5 seconds, then pauses for 5 seconds, then repeats the cycle, how many times does he slap his tail on the water in one minute?

Solution: If the beaver slaps the water 4 times in five seconds, and then doesn't slap it for five more seconds, the beaver is slapping the water 4 times every 10 seconds.

Next, find how many times this cycle happens in a minute. Since 1 minute is 60 seconds, the beaver repeats this cycle $60 \div 10 = 6$ times in a minute. Finally, if one cycle has 4 slaps, the beaver must slap the water a total of $4 \times 6 = \boxed{24 \text{ times}}$ in a minute.

29. If the lengths of two of the sides of an isosceles triangle are 6 and 12, then what is the perimeter of this triangle?

    Solution: First, define an isosceles triangle. An isosceles triangle has two sides of equal length.

    In this problem, there are two options. The first option is that there are two sides of the triangle that have lengths of 6. The second option is that there are two sides which both have lengths of 12.

    However, in the first option, if there is one side which is 12, then the two sides which are 6 would form a straight line next to the line with length 12 because $6 + 6 = 12$.

    Therefore, the triangle must have two sides which are 12 and one side which is 6. Now, add up all the sides to find the perimeter: $12 + 12 + 6 = 30$. Therefore, the triangle has a perimeter of $\boxed{30}$.

30. If there are 300 cubbyholes and Joe has 301 flyers, what is the probability that one cubby will have at least two flyers?

    Solution: If each cubbyhole is filled with one flyer, than that will leave $301 - 300 = 1$ flyer left over. However, since all cubbies are occupied, putting the extra flyer in another cubby will cause at least one cubby to have two flyers. This means that the probability that one cubby will have at least two flyers is $\boxed{100\% \text{ or } 1}$.

Target Round

1. On the planet Tums, someone has a stomach ache every 30 seconds. How many stomach aches occur in 42 minutes?

   Solution: First, remember that there are 60 seconds in one minute. Then, calculate how many stomach aches there are in one minute. Since there are 60 seconds in a minute, and a stomach ache happens every 30 seconds, there must be $60 \div 30 = 2$ people that get a stomach ache every minute.

   Next, calculate how many people get stomach aches in 42 minutes. As mentioned above, two people get a stomach ache every minute. To solve this, multiply the amount of minutes (42) by the stomach aches per minute (2) to discover how many stomach aches there are in 42 minutes. This means there are $42 \times 2 =$ $\boxed{\text{84 stomach aches}}$ on the planet Tums in 42 minutes.

2. I have four nickels, five quarters, and 6 dollar bills. How much money do I have left if I spend \$1.50 on lollipops? Express your answer as a decimal rounded to the nearest hundredth.

   Solution: Start by determining the value of each coin. Nickels are worth 5 cents, quarters are worth 25 cents and dollar bills are worth one dollar. Also, remember there are 100 cents in each dollar.

   Next, convert each coin to a dollar value. Since nickels are worth 5 cents and there are 100 cents in a dollar, nickels are worth $5/100$ dollars or \$0.05. Also, since quarters are 25 cents, each quarter is worth $25/100$ dollars or \$0.25.

   Now, find how many dollars you have in total. First, having four nickels at \$0.05 each will give a total of $\$0.05 \times 4 = \$0.20$ in nickles. Next, five quarters worth \$.25 each will give a total of $\$0.25 \times 5 = \$1.25$ in quarters. Next, add all of the money together. This gives the total value of $\$6.00 + \$0.20 + \$1.25 = \$7.45$.

   Lastly, subtract the \$1.50 form the total \$7.45 to find how much money is left over. This results in the final answer: $\$7.45 - \$1.50 =$ $\boxed{\$5.95}$.

3. How much wood (in logs) could a woodchuck chuck (in 35 minutes) if a woodchuck could chuck wood (at a rate of 36 logs per hour)?

   Solution: For this problem, it helps to first simplify the question, and get rid of any extra information. In this question, the goal is to find how many logs of wood a woodchuck could chuck in 35 minutes, if it is working at a pace of 36 logs per hour.

Since the woodchuck chucks 36 logs per hour, and there are 60 minutes in an hour, the woodchuck chucks 36 logs every 60 minutes. However, the question is asking how much wood could a woodchuck chuck in 35 minutes. Let that amount be w.

Then, use a proportion to solve this problem. If a woodchuck chucks 36 logs per 60 minutes, and it chucks w logs in 35 minutes, this gives the proportion

$$\frac{36\ logs}{60\ min} = \frac{w\ logs}{35\ min}$$

Solving this proportion shows that $w = \frac{36 \times 35}{60} = \frac{6 \times 3 \times 2 \times 5 \times 7}{6 \times 5 \times 2} = 3 \times 7 = 21$, meaning that a woodchuck can chuck $\boxed{\text{21 logs}}$ in 35 minutes.

4. Harry receives a \$50 parking ticket for parking his broomstick in a red zone. Because he's Harry, he gets a 20% discount. How much does Harry have to pay (in dollars)?

   Solution: Since Harry gets a 20% discount, he has to pay $100\% - 20\% = 80\%$ of the normal ticket price. To find the amount of money Harry has to pay, first convert the percentage to a fraction: $80\% = \frac{80}{100} = \frac{4}{5}$. Next, multiply $4/5$ by the normal ticket price of \$50. This gives the result of $\frac{4}{5} \times \$50 = \$40$. That means Harry must pay $\boxed{\$40}$ for his parking ticket.

5. Danielle is buying lunch at school. She has a choice of a sandwich, popcorn chicken, or a burger plus one of 4 drinks, one of 3 fruits, and one of 5 vegetables. How many choices does she have for her lunch if she gets one of each type?

   Solution: For lunch, Danielle can chose 1 out of 3 main dishes, 1 out of 4 drinks, 1 out of 3 fruits and 1 out of 5 vegetables. Then, in order to find how many different combinations of lunches Danielle can choose, multiply the amount of choices for each option to find how many total possibilities there are. This means there are $3 \times 4 \times 3 \times 5 = 12 \times 15 = \boxed{\text{180 choices}}$ that Danielle can choose from for lunch.

6. Sarah is delivering newspapers on a street of houses numbered 1 to 100 with even-numbered houses on the right and odd-numbered houses on the left. If she only delivers to house numbers that are a multiple of 3 and on the right, how many houses does she deliver newspapers to?

   Solution: First, realize that since all numbers on the right are even, then all of these numbers are multiples of 2. Sarah only delivers newspapers to houses that are a multiple of 3 and that are on the right, so that means she delivers newspapers to houses that are a multiple of 3 and of 2.

   In order to be a multiple of 3 and a multiple of 2, the number has to be a multiple of $3 \times 2 = 6$. Therefore, Sarah delivered the newspapers to all houses with numbers that are a multiple of 6. To find the number of houses she delivered to, divide the number of houses (100) by 6 to find the total number of multiples of 6. This results in $100 \div 6 = 16\frac{4}{6}$, but the remainder does not give another multiple of 6, so it can be ignored. Therefore, she delivered newspapers to $\boxed{\text{16 houses}}$.

7. A room is one-half full of people. After 20 people leave, the room is one-third full. How many people would be in the room if it were filled?

   Solution: Let c be the number of people in the room if it is full.

   At the start the problem, there are $\frac{1}{2}c$ people because the room is one-half full of people. Then, after 20 people leave, the room is one-third full, which results in the equation $\frac{1}{2}c - 20 = \frac{1}{3}c$.

   Next, isolate the variable by putting both of the terms with variables on one side: $20 = \frac{1}{2}c - \frac{1}{3}c$. To subtract fractions, create a common denominator and simplify the equation: $20 = \frac{3}{6}c - \frac{2}{6}c = \frac{1}{6}c$. Finally, solve for c by multiplying both sides of the equation $\frac{1}{6}c = 20$ by 6 to get $c = 120$. This means the full room capacity is $\boxed{\text{120 people}}$.

8. Paul is not very good at talking to girls. Every time Paul tries to talk to a girl, he gets very nervous and starts counting in multiples of 3. If Paul counts in multiples of 3 (3, 6, 9, 12, 15,...) until he reaches 1,242, how many numbers has Paul said?

   Solution: In this problem, Paul always follows a consistent pattern, and only says multiples of 3. This is because the first number of the sequence is $3 = 1 \times 3$, the second number of the sequence is $6 = 2 \times 3$, the third number is $9 = 3 \times 3$, and so on.

   This means that the nth number in the sequence is $n \times 3$. Therefore, to find out how many numbers Paul has said, set $n \times 3$ equal to 1,242. Then, after 1,242 dividing by 3, it is revealed that 1,242 is the 414th number he has said. This means Paul has said a total of $\boxed{\text{414 numbers}}$.

3rd-4th Grade 2015 Solutions

Sprint Round

---

1. What is $2015 \times 20 \times 15 \times 2 \times 0 \times 1 \times 5$?

   Solution: Although this problem may seem very difficult as there are many numbers to multiply out, it is actually quite simple. Before solving anything, analyze the ENTIRE problem. As you can see, there is a 0 being multiplied in the equation. Nothing needs to be multiplied because any number that is multiplied by 0 equals $\boxed{0}$.

2. How many feet is 26 yards?

   Solution: In order to solve this, know that 3 feet is equivalent to 1 yard. To find how many feet are in 26 yards, set up an equation that can convert yards to feet. To do this, set up this equation:

   $$26\ yards \times \frac{3\ feet}{1\ yard} = \boxed{78\text{ feet}}$$

3. If I have 205 pennies and exchange them for some number of nickels, how many nickels would I have?

   Solution: Since 5 pennies are equal to 1 nickel, this problem can be solved by setting up a similar equation to Problem 2, but instead pennies are converted to nickels.

   $$205\ pennies \times \frac{1\ nickel}{5\ pennies} = \boxed{41\text{ nickels}}$$

4. It is currently 2:08 PM. How many minutes will pass before it is 4:01 PM (of the same day)?

   Solution: Right now, it is 2:08 PM. To start, find how many minutes until it is 3:00 PM. Since there are 60 minutes in 1 hour, the amount of time until 3:00 PM is $60 - 8$ minutes, which equals 52.

   So, if it is 3:00 PM, there will be 60 minutes until it is 4:00 PM. If it is 4:00 PM, there will be 1 minute until it is 4:01 PM. So, from 2:08 PM to 4:01 PM, $52 + 60 + 1$ minutes will pass, which equals a total of $\boxed{113\text{ minutes}}$.

5. The area of a square is 49 square feet. What is the length of the side of the square (in feet)?

   Solution: The area of a square is $A = s \times s = s^2$, where A is the area, and s is the length of a side of the square. So, plugging in 49 square feet into A, the equation becomes $49\text{ ft}^2 = s^2$. Taking the square root of 49 shows the side length of the square is $\boxed{7\text{ ft}}$.

6. The perimeter of an equilateral triangle is the same as that of a square. If the side length of the square is 12, what is the side length of the triangle?

   Solution: Realize that an equilateral triangle is a triangle where all sides are congruent, or have the same length. Therefore, the perimeter of the triangle is $P = 3 \times t$, where P is the perimeter of the triangle, and t is a side of the triangle.

   In addition, the triangle's perimeter is equal to the perimeter of a square, whose side is equal to 12. The perimeter of a square is $4 \times s$, where s is a side of the square. By plugging 12 into this equation, the perimeter becomes $4 \times 12 = 48$. So, because the perimeter of the triangle is equivalent to the perimeter of the square, plug 48 into the equation for the perimeter of the triangle, giving the equation $48 = 3 \times t$. Dividing both sides of the equation by 3 gives the length of the side of the triangle: $t = \boxed{16}$.

7. If $a||b = \frac{a+b}{ab}$, what is $4 \,||\, (5 \,||\, 6)$? Express your answer as a common fraction.

   Solution: To start, understand that $a||b = \frac{a+b}{ab}$. Then, to solve for $4 \,||\, (5 \,||\, 6)$, follow the order of operations, and solve what is within the parentheses first. So, plugging in $a = 5$ and $b = 6$ gives the result of:

$$\frac{5+6}{5 \times 6} = \frac{11}{30}$$

   Now, the new equation to solve is $4||\frac{11}{30}$. Plugging in $a = 4$ and $b = \frac{11}{30}$ into the new equation gives the result of:

$$\frac{4+\frac{11}{30}}{4 \times \frac{11}{30}}$$

   Then, simplify this expression into a common fraction. For the numerator, turn 4 into a fraction with a denominator of 30, which makes the numerator $\frac{120}{30} + \frac{11}{30} = \frac{131}{30}$ as our numerator. As for the denominator, $4 \times \frac{11}{30} = \frac{22}{15}$. Then, realize that dividing $\frac{131}{30}$ by $\frac{22}{15}$ is the same as multiplying the numerator by the reciprocal of the denominator. This gives the equation:

$$\frac{131}{30} \div \frac{22}{15} = \frac{131}{30} \times \frac{15}{22} = \frac{131}{2} \times \frac{1}{22} = \boxed{\frac{131}{44}}$$

8. Matthew Tang and Arthur Liu are playing a card game in their free time. If each card game takes 160 seconds, how many games are they going to play in 2 hours?

Solution: Realize that there are 60 seconds in one minute, and 60 minutes in one hour. Using this, convert 2 hours into seconds by using these ratios of time, like so:

$$2\ hr \times \frac{60\ min}{1\ hr} \times \frac{60\ sec}{1\ min} = 7,200\ seconds$$

Then, it is easy to figure out how many 160-second games that can be played. Do this by multiplying the total amount of time Arthur and Matthew are going to play (7,200 seconds) by the ratio of time per game previously stated in the problem (1 game every 160 seconds). So,

$$7,200\ sec \times \frac{1\ game}{160\ sec} = \frac{720}{16} = \boxed{45 \text{ games}}$$

9. If there are 5 quarks in a quack, 3 quacks in a quat, and 7 quats in a quirt, how many quarks are in 20 quirts?

   Solution: In this problem, use the ratios of the quarks, quacks, quats, and quirts to convert from one to another. In each fraction that is multiplied, the denominator cancels out the numerator of the fraction before it, therefore converting from one unit to another. Since the amount of quirts is given, convert quirts to quats, then quats to quacks, and then quacks to quarks to get the final answer. This process creates the equation:

$$20\ quirts \times \frac{7\ quats}{1\ quirt} \times \frac{3\ quacks}{1\ quat} \times \frac{5\ quarks}{1\ quack} = \boxed{2,100\ quarks}$$

10. A 23-page packet contains a total of 3450 words. On average, how many words are on each page?

    Solution: The average amount of words per page can be calculated by dividing the total word count by the total page count. In this problem, divide 3450 by 23 pages to get the average words per page:

$$\frac{3450\ words}{23\ pages} = \boxed{150\frac{words}{page}}$$

11. A person walks into the airport every 5 seconds. How many people walk into the airport in 20 minutes?

    Solution: First, convert 20 minutes into seconds. Since there are 60 seconds in one minute, use the ratio $\frac{60\ sec}{1\ min}$ to perform this operation.

$$20\ min \times \frac{60\ sec}{1\ min} = 1,200\ sec$$

Then, since 20 minutes is equivalent to 1,200 seconds, it is easy to find how many people walk into the airport using the ratio of $\frac{1\ person}{5\ sec}$.

$$1,200\ sec \times \frac{1\ person}{5\ sec} = 240\ people$$

Therefore, $\boxed{\text{240 people}}$ walk into the airport every 20 minutes.

12. What is the least number of coins I need to make 37 cents?

    Solution: To find the least number of coins to make 37 cents, choose larger coins first. So, the largest coin is the quarter, worth 25 cents. Since 37 is larger than 25, more coins are needed, but 2 quarters equals 50 cents, so a smaller coin must be chosen.

    The next coin is the dime, worth 10 cents. So, the total becomes $25 + 10 = 35$, which is still less than 37, but another dime cannot be chosen. The next coin, the nickel is worth 5 cents, so this coin cannot be used either.

    So, all that's left is the penny, worth 1 cent each, and since 2 more cents are still needed, 2 pennies must be chosen. Adding up the number of coins used, the result is 1 quarter + 1 dime + 2 pennies = $\boxed{\text{4 coins}}$.

13. Person A, Person B, and Person C have 54 dollars each. Person A uses $\frac{1}{2}$ of their money, Person B uses $\frac{2}{3}$ of their money, and Person C uses $\frac{2}{9}$ of their money to buy food. How many dollars do they have left altogether?

    Solution: If Person A used $\frac{1}{2}$ of their money, then the money they have left is

    $$54\ dollars \times \frac{1}{2} = 27\ dollars$$

    If Person B used $\frac{2}{3}$ of their money, then the money they have left is

    $$54\ dollars \times \frac{1}{3} = 18\ dollars$$

    left. If Person C used $\frac{2}{9}$ of their money, then the money they have left is

    $$54\ dollars \times \frac{7}{9} = 42\ dollars$$

    Adding all the money each person had left gives a total amount of

    $$\$27 + \$18 + \$42 = \boxed{\$87}$$

14. Compute the sum of the first 11 even natural numbers.

    Solution: In order to solve this problem, know that natural numbers are the positive integers (1, 2, 3...). So, the first 11 even natural numbers are 2, 4, 6, 8, 10, 12, 14, 16, 18, 20, and 22. Adding these numbers together equals 132.

    Tip: To find the sum of these numbers faster, realize that the sum of 2 and 22, 4 and 20, 6 and 18, 8, and 16, and 10 and 14 all equal

24. This can make the addition a simple multiplication problem, as shown below:

$$\begin{aligned} &2+4+6+8+10+12+14+16+18+20+22 \\ =&(2+22)+(4+20)+(6+18)+(8+16)+(10+14)+12 \\ =&24+24+24+24+24+12=(24\times 5)+12=\boxed{132} \end{aligned}$$

15. What is the value of $7 \times (2+3-4\times 2)$?

Solution: Start with looking back at the rules of the order of operations, or PEMDAS (**P**arentheses, **E**xponents **M**ultiplication, **D**ivision, **A**ddition, and **S**ubtraction). So, to solve $7 \times (2+3-4\times 2)$, start with what is within the parentheses.

Inside, there are no exponents, so move to multiplication: 4×2. $4\times 2 = 8$, which leaves (2+3-8). Solving for what is in the parentheses results in $2+3-8=-3$.

After solving for the equation within the parentheses, simply multiply 7 and -3 to get the final answer: $7 \times -3 = \boxed{-21}$.

16. Train A leaves New York and travels north at 80 mph. Train B leaves New York at the same time but travels south at 100 mph. After 3 hours, how many miles apart are the two trains?

Solution: To solve this problem, it is helpful to know that $distance = rate \times time$, or $d = r \times t$. Train A travels north at 80 mph for 3 hours, so the distance it travels is $d = 80 \times 3 = 240$ miles. Train B travels south at 100 mph for 3 hours, so the distance it travels is $d = 100 \times 3 = 300$ miles. Since the trains traveled in opposite directions, the total distance between them is $240 + 300 = \boxed{\text{540 miles}}$.

17. If today is Saturday, what day was it 100 days ago?

Solution: To solve this, divide 100 by 7 to find how many whole weeks have passed, and how many "extra" days led up to this Saturday. So, $100/7 = 14$ with a remainder of 2. Since there is a remainder of 2, count back 2 days from Saturday to find the answer, which is $\boxed{\text{Thursday}}$.

18. What is the probability that a randomly chosen year in the interval 2000 to 2100 (inclusive) will be a leap year? Express your answer in a common fraction. (Leap years are those that are a multiple of 400 or a multiple of 4 but not 100.)

Solution: To count the number of leap years, first start with the years that are multiples of 400. From 2000 to 2100 inclusive, there is 1 multiple of 400, which is 2000.

Next, use the rule that a leap year is a multiple of 4 but not 100. To do this, first find how many numbers are divisible by 4, and exclude the number that is a multiple of 100 (which is clearly 2100). To find the amount of numbers divisible by 4, we divide 100 by 4, which gives a total of 25 years. However, by excluding the year 2100, there are $25-1=24$ leap years that are divisible by 4 and not 100. Adding the one leap year that is divisible by 400 gives a final total of $24+1=25$ leap years.

Then, in order to find the probability, take the number of leap years, and divide it by the number of years that can be chosen from. To find how many years there are between two numbers, subtract the bigger number from the smaller number, and add one. This means there are $2100-2000+1=101$ years between 2000 and 2100 inclusive, so the odds of choosing a leap year are $\boxed{\frac{25}{101}}$.

19. You have a rectangle, and if you increase the length by 10% and increase the width by 5%, by what percent is the area increased by? Express your answer as a decimal rounded to the nearest tenth.

    Solution: First, realize that the area of a rectangle is equal to the length multiplied by the width ($Area = l \times w$).

    By increasing the length by 10%, it means 10% of the length is added to the original length. Therefore, the new length is:

    $$(100+10)\% \times l = 110\% \times l = 1.1 \times l$$

    Using the same method, increasing the width by 5% makes the new width:

    $$(100+5)\% \times w = 105\% \times w = 1.05 \times w$$

    With the new width and length, the new area of the rectangle is:

    $$1.1\ l \times 1.05w = 1.155 \times l \times w = 1.155 \times Area = 115.5\% \times Area$$

    If the area of the new rectangle is 115.5% of the original area, the area must have increased by $\boxed{15.5\%}$.

20. Bill has 5 unique Rubik's Cubes to give away to 4 people: Leon, Suhn, Mitchell, and Bill 2.0. How many ways can he distribute them?

    Solution: Since each Rubik's Cube can be given to one of 4 people, there must be 4 ways to hand out each Rubik's cube. Since there are 5 unique Rubik's cubes, and each cube can be handed out 4 ways, there are $4 \times 4 \times 4 \times 4 \times 4$ ways to give out all five Rubik's cubes. $4 \times 4 \times 4 \times 4 \times 4 = 4^5 = 1024$, so there are $\boxed{1024 \text{ ways}}$ to distribute the Rubik's cubes.

21. What is the positive difference between $1+2+3+4+5+\ldots+100$ and $2+3+4+5+6+\ldots+101$?

Solution: All the numbers in the first set of numbers are also in the second set of numbers, except for the number 1, and all the numbers in the second set of numbers are in the first set of numbers except 101. So, the difference between the 2 sets of numbers is the difference between 101 and 1. So our answer is 101-1=100.

Solution 2: Let $a = 2+3+4+5+\ldots+100$. This means that the first set of numbers is $1+a$, and the second set of numbers is $a+101$. Then, subtract $a+1$ from $a+101$:

$$a + 101 - (a + 1) = 101 - 1 = \boxed{100}$$

22. A sequence is defined by $a_n = a_{n-1} \times n$. If a_5 is equal to 180, find the value of a_2.

Solution: In order to solve the problem, work backwards and find a_4, then a_3, and finally a_2.

To find a_4, plug $n = 5$ into the equation, which results in $a_5 = a_4 \times 5$. After substituting a_5 for 180 and then simplifying, $a_4 = 36$. Next, plug $n = 4$ into the same equation to get a_3.

Plugging $n = 4$ into the equation results in $a_4 = a_3 \times 4$. After substituting a_4 for 36 and solving, $a_3 = 9$. Then, plug in $n = 3$ to solve for a_2.

Plugging $n = 3$ into the equation results in $a_3 = a_2 \times 3$, and after substituting a_3 for 9 and solving results in $a_2 = \boxed{3}$.

23. The sides of a right triangle have lengths 8 cm, x cm, and 15 cm, and x is an integer. What is the area of the triangle (in square centimeters)?

Solution: In order to solve this problem, use the Pythagorean Theorem. The Pythagorean Theorem states that for each right triangle $a^2 + b^2 = c^2$, where a and b are the lengths of the two legs, and c is the length of the hypotenuse.

Since x could be less than or greater than 15, it could be the leg or the hypotenuse. Therefore, there are two cases:

Case 1: x is a leg of the triangle

$$\begin{aligned} 8^2 + x^2 &= 15^2 \\ x^2 &= 225 - 64 \\ x &= \sqrt{161} \end{aligned}$$

In this case, x would not be an integer, so x cannot be a leg of the triangle.

Case 2: x is the hypotenuse

$$8^2 + 15^2 = x^2$$
$$x^2 = 225 + 64$$
$$x = \sqrt{289} = 17$$

In this case, x would be an integer, so x must be the hypotenuse of the triangle.

For a right triangle, the area is $\frac{1}{2} \times \textit{base} \times \textit{height}$, which is equal to $\frac{1}{2}$ times the product of the legs of the triangle. Since x is the hypotenuse, the area of the triangle is $\frac{1}{2} \times 8 \times 15 = 4 \times 15 = \boxed{60}$.

24. In triangle ABC, angle A is equal to 65 and angle B is equal to 75. The bisector of angle C is drawn to meet AB at point D. What is the measure of angle ADC? (An angle bisector is a line that cuts an angle into two angles of equal measure.)

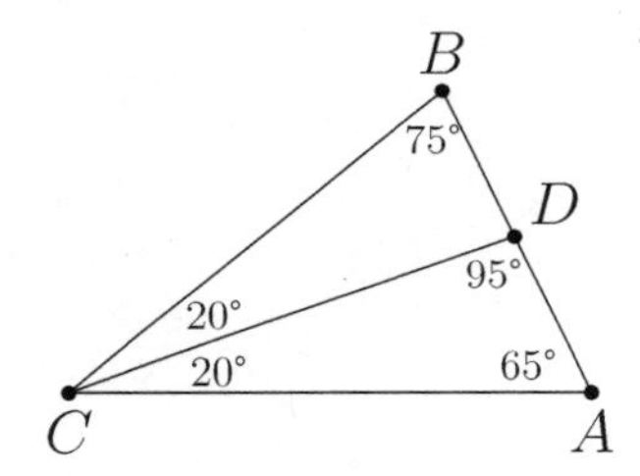

Solution: Before attempting this problem, it helps to know that the sum of the angles in a triangle is equal to 180°. Therefore, find angle C by subtracting 65° and 75° from 180°. This means angle C is equal to:

$$180° - 75° - 65° = 40°$$

If a bisector is drawn through angle C, that means the bisector divides angle C into two equal angles that are each half of angle C, which is $40° \div 2 = 20°$. So, if angle A is equal to 65°, and angle ACD is equal to 20°, then angle ADC must equal $180° - 65° - 20° = \boxed{95°}$.

25. Five friends, Natalie, Lucy, Amy, Thomas, and Ian, go to Disneyland together. They each buy their own ticket for \$156 each. They equally split up the lunch bill of \$100 and they each buy one pair of Mickey Mouse ears for \$22 a pair. If this is all they spent, how many dollars did each of them spend at Disneyland?

Solution: Each of their tickets cost \$156. If they split up the lunch bill among the 5 of them, then they each spent $\$100 \div 5 = \$20$ on lunch. And finally, they bought a pair of Mickey Mouse ears for \$22. So, each person spent a total of $\$156 + \$20 + \$22 = \boxed{\$198}$.

26. William, Zack, Bryan, Jason, and Joey are running in a race. These statements are true about the results:

    William finished before Jason but after Joey.

    Jason finished in third place.

    Bryan finished after Joey but he did not finish last.

    Zack finished after both Bryan and Jason.

    Who was the first place runner?

    Solution: With these types of questions, it helps to draw a diagram similar to the one below. Imagine that they are in a line, and the person on the left side of the line is the one who finished first, and the person on the right side of the line is the one who finished last. The most specific piece of information that is given is that Jason finished in third place. This means he can be placed in the middle:

    ______ ______ Jason ______ ______

    Next, William finished before Jason, but after Joey. Since there are only two spots in front of Jason, Joey must have been the first to finish, followed by William:

    Joey William Jason ______ ______

    Next, Zack finished after Bryan and Jason. Because there are only two spots left after Jason, this must mean Zack finished last, and Bryan finished second to last:

    Joey William Jason Bryan Zack

    Therefore, $\boxed{\text{Joey}}$ was the first place runner. (Note: The last paragraph of information is not necessary to find the answer to the problem, but it helps complete the entire diagram.)

27. If the lengths of two sides of a isosceles triangle are 5 and 8 respectively, what is the sum of all the possible perimeters for the triangle?

    Solution: An isosceles triangle is a triangle with two sides that are congruent, or equal. So, since two of the triangle's sides are 5 and 8, then there are two possible triangles: one with sides 5, 5, and 8, and the other with sides 5, 8, and 8.

    But before assuming those as the answers, make sure that triangles with those sides are possible. Within a triangle, the sum of two sides must always be greater than the third remaining side. Luckily, $5+5>8$ and $5+8>8$, so both triangles are possible solutions.

The perimeter for the first possible triangle equals $5+5+8=18$ and the perimeter for the second possible triangle equals $5+8+8=21$. Taking the sum of these two triangles gives the final answer of $18+21=\boxed{39}$.

28. Bob sells gum in packs of 3 and 5. What is the largest amount of gum that a person is not able to buy using a combination of Bob's packs of gum? For example, a person can get 14 pieces of gum by ordering 3 packs of 3 and 1 pack of 5.

    Solution: Let n represent any non-negative integer. First off, since the gum is sold in packs of 3, a person can buy $3n$ (meaning any multiple of 3) sticks of gum. Also, they can buy $5+3n$ sticks of gum by buying one pack of 5 and n packs of 3. In addition, they can buy $10+3n$ sticks of gum by buying two packs of 5 and n packs of 3.

    Notice that for any number greater than 10, a person can buy that many pieces of gum by using one of the three methods above. This means that the greatest amount of pieces someone cannot buy must be less than 10. Then, plug in numbers starting from 9. It is possible to get 9 pieces using 3 packs of 3, and to get 8 pieces using a pack of 3 and a pack of 5. However, it is not possible to get 7 pieces of gum. Therefore, the answer is $\boxed{\text{7 pieces of gum}}$.

29. Susan is a criminal case solver. Today, her case is to find out who ate which type of candy, Skittles, M&M's, Jolly Ranchers, and Twix. She knows Michael hates chocolate. Mindy says she would never eat Twix in a million years. John said that he saw a boy eat Skittles. Lastly, Joey is under a lot of pressure, and confesses that he ate the Jolly Ranchers. Given that nobody ate the same candy, what did Mindy eat? (Assume M&M's and Twix are the only two chocolate candies.)

    Solution: To start off, list out all of the suspects who ate candy: Joey, Michael, Mindy, and John. Joey must have eaten Jolly Ranchers because of his confession. Also, John said he saw a boy eat Skittles, and since Michael hates chocolate and did not eat the Jolly Ranchers, then Michael must have eaten the Skittles. Now, the people left are Mindy and John who each ate either M&M's and Twix. Mindy said she would never eat Twix in a million years, so she must have eaten the $\boxed{\text{M\&M's}}$.

30. Using a standard deck of cards, a class is doing a fun game. When a certain suit is a pulled from the deck, the class has to do a certain exercise. Hearts are jumping jacks, diamonds are sit-ups, clubs are push-ups, and spades are burpies. If all face cards (including aces)

represent 10, and the number on the card represent how many of each exercise you need to do, how many sit-ups and push-ups would a student have completed after going through the whole deck?

Solution: Realize that there are 52 cards in a standard deck, and there are 13 of each suit: hearts, diamonds, clubs, and spades. Note that the face cards (King, Queen, Jack and Ace) each mean the class must do 10 of a particular exercise. So, the number of sit-ups and push-ups a student needs to do is two times the sum of one suit of cards. The sum of one suit equals $2+3+4+5+6+7+8+9+10+(4 \times 10)$, which equals 94. So, the total number of sit-ups and push-ups equals $2 \times 94 = \boxed{188}$.

Target Round

---

1. Compute $\frac{25\times7+2}{3}$.

   Solution: First, start by computing the numerator: $25 \times 7 + 2$. To do this, follow the rules of the order of operations, or PEMDAS (**P**arentheses, **E**xponents **M**ultiplication, **D**ivision, **A**ddition, and **S**ubtraction). This says to multiply 25×7 first, which equals 175. Then, finish computing the numerator by adding the 2: 175+2=177. Finally, divide the numerator by the denominator: $\frac{177}{3} = \boxed{59}$.

2. Jenna has 10 pieces of candy. She gives seven pieces to her brother. Then her friend gives her 2 pieces of candy. How many pieces of candy does she have now?

   Solution: She starts with 10 pieces of candy. If she gives her brother 7 pieces, she has $10 - 7 = 3$ pieces of candy left. Then, if her friend gives her 2 more pieces of candy, then she ends up with $3 + 2 = \boxed{\text{5 pieces of candy}}$.

3. How many cents is 26 pennies, 3 nickels, and 5 dimes?

   Solution: A penny is worth 1 cent, so if there are 26 pennies, they equal $26 \times 1 = 26$ cents of pennies. A nickel is worth 5 cents, so if there are 3 nickels, they equal $3 \times 5 = 15$ cents of nickels. Finally, a dime is worth 10 cents. If there are 5 dimes, they equal $5 \times 10 = 50$ cents of dimes. Adding these together gives a result of $26 + 15 + 50 = \boxed{\text{91 cents}}$.

4. What is the area of a rectangle with side lengths 5 and 7?

   Solution: The area of a rectangle is equal to the length of one side times the length of the other side ($Area = length \times width$). If the side lengths are 5 and 7, then the area equals $5 \times 7 = \boxed{35}$.

5. What is the result when the fifth even natural number is multiplied by the fifth odd natural number?

   Solution: First, note that natural numbers are the positive integers (1, 2, 3...). Then, list out the first five even natural numbers, like so: 2, 4, 6, 8, 10. This means 10 is the fifth even natural number. Then, listing out the first five odd natural numbers results in this list:
   1, 3, 5, 7, 9. So, 9 is the fifth odd natural number. Then, multiplying the fifth even natural number, 10, by the fifth odd natural number, 9, results in $10 \times 9 = \boxed{90}$.

6. A strange clock chimes 1 time at 1 o' clock, 2 times at 2 o'clock, 3 times at 3 o' clock, 4 times at 4 o' clock, 5 times at 5 o' clock, and so on. How many times will the clock chime from 1-12 o' clock?

   Solution: If the clock chimes 1 time at 1 o' clock, 2 times at 2 o' clock, 3 times at 3 o' clock and so on, then the number of times the clock will chime from 1-12 o'clock equals:

$$\begin{aligned} &1 + 2 + 3 + 4 + 5 + 6 + 7 + 8 + 9 + 10 + 11 + 12 \\ &= (1 + 2 + 3 + 4) + (5 + 6 + 7) + (8 + 12) + (9 + 11) + 10 \\ &= 10 + 18 + 20 + 20 + 10 \\ &= \boxed{78 \text{ times}} \end{aligned}$$

7. The perimeter of a square is 5 more than the perimeter of a triangle with side lengths 4, 5, and 6. What is the area of the square?

   Solution: First, start off with the perimeter of the triangle. If the triangle has side lengths 4, 5, and 6, then the perimeter of the triangle equals $4+5+6 = 15$. If the perimeter of the square is 5 more than the perimeter of the triangle, then the perimeter of the square is equal to $15 + 5 = 20$.

   Since a square has 4 sides of equal length, each side has a length of $20 \div 4 = 5$. The area of a square is $A = s \times s$, where s is the length of a side, so the area of the square is equal to $5 \times 5 = \boxed{25}$.

8. How many integers are there between 1 and 100 (inclusive) that can be expressed as the sum of three positive consecutive integers?

   Solution: In order to solve this problem, it is important to know that three consecutive numbers means three numbers that are next to one another on the number line (Ex. 5, 6, and 7 would be consecutive numbers, but 5, 7, and 9 would not.)

   Then, realize the first set of positive consecutive numbers that can be added to get a number between 1 and 100 is $1 + 2 + 3=6$.

   Then, divide 100 by 3, and get 33 with a remainder of 1. Next, realize that if adding $33 + 33 + 33$ gets 99, then adding $32 + 33 + 34$ will also get 99. This is the last set of 3 consecutive numbers that can be added together to express an integer between 1 and 100.

   That means every set of three consecutive numbers that starts with a number between 1 to 32 (including 1 and 32) will express a number between 1 and 100. That means there are $\boxed{\text{32 integers}}$ that can be expressed as the sum as three consecutive integers.

3rd-4th Grade 2016 Solutions
Sprint Round

---

1. You had 20 apples. You ate half of your apples. Your friend gave you 6 more apples. How many apples do you have now?

   Solution: At the beginning, you started with 20 apples. Then, if you ate half of your apples, you have eaten $20 \div 2 = 10$ apples. If you have eaten 10 out of 20 apples, you had $20 - 10 = 10$ apples remaining. Finally, once your friend gave you 6 more apples, your final amount of apples is $10 + 6 = \boxed{16 \text{ apples}}$.

2. What is the area of a triangle with height 7 and base 8?

   Solution: The area of a triangle is $\frac{1}{2} \times base \times height$. Since the base of the triangle is 8, and the height of the triangle is 7, the area of the triangle must be $\frac{1}{2} \times 8 \times 7 = 4 \times 7 = \boxed{28}$.

3. The perimeter of a square with side length 6 is twice the perimeter of an equilateral triangle. What is the side length of the triangle?

   Solution: First, find the perimeter of the square. Since there are 4 sides of a square, and each side has a length of 6, the perimeter is $4 \times 6 = 24$.

   Then, since the perimeter of the square is twice the perimeter of the triangle, the perimeter of the triangle must be half of the perimeter of the square. This means the perimeter of the triangle is $24 \div 2 = 12$.

   Finally, an equilateral triangle has three sides of equal length, so the side length of the triangle is $12 \div 3 = \boxed{4}$.

4. Maya thinks of a number, multiplies it by 7, adds 14, and says the result out loud. If Maya says "84", what number was she originally thinking of?

   Solution: Let N be the number Maya was originally thinking of. Since the number is multiplied by 7, then 14 is added to it, the new number becomes $(N \times 7) + 14$. Since the new number equals 84, $(N \times 7) + 14 = 84$.

   Subtracting 14 from each side of the equation shows that $N \times 7 = 70$. Dividing by 7 on each side of the equation then gives the final answer: $N = \boxed{10}$.

5. Clarisse flips a coin twice. What is the probability that the two flips yield different results?

Solution: There are four possible results when flipping a coin twice: two heads, a head and a tail, a tail and a head, or two tails. 2 of the 4 possibilities give different results, so the answer is $\frac{2}{4}$, or $\boxed{\frac{1}{2}}$.

6. A regular pentagon has one side with the value 6. What is the total perimeter?

   Solution: A regular pentagon has five sides that all have the same length. Since one side has a length of 6, the total perimeter is:

$$5 \times 6 = \boxed{30}$$

7. Suppose that $a \diamond b = \frac{a \times b}{a+b}$. Compute $2 \diamond (3 \diamond 6)$.

   Solution: First, realize that according to the order of operations, operations in parentheses must be solved first. This means that in the equation $2 \diamond (3 \diamond 6)$, $(3 \diamond 6)$ must be solved first.

$$2 \diamond (3 \diamond 6) = 2 \diamond (\frac{3 \times 6}{3+6}) = 2 \diamond 2$$
$$2 \diamond 2 = \frac{2 \times 2}{2+2} = \boxed{1}$$

8. Find the number of divisors of 24, including 1 and 24.

   Solution: A divisor is a number that can divide into 24 without leaving a remainder. Using division, quickly find that the divisors of 24 are 1, 2, 3, 4, 6, 8, 12, and 24. Counting them up, this gives a total of $\boxed{\text{8 divisors}}$.

9. Thirteen 2 by 2 squares are pasted together in a straight line. What is the perimeter of the resulting figure?

   Solution: The resulting figure is a long, narrow rectangle, as shown below. The length of this rectangle is the side length of all the squares, which is $2 \times 13 = 26$. The width is the side length of one square, which is 2.

   To find the perimeter, remember the formula for the perimeter of a rectangle is $2 \times (length + width)$. This makes the perimeter of the figure $2 \times (26 + 2) = 2 \times 28 = \boxed{56}$.

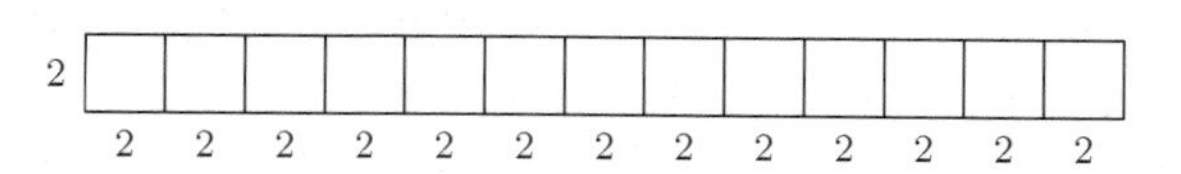

10. Allison has to solve 15 math problems for homework. 5 are easy problems, 5 are medium problems, and 5 are hard problems. If it takes her 2 minute to solve an easy problem, 4 minutes to solve a

medium problem, and 7 minutes to solve a hard problem, how many minutes does it take her to finish her homework?

Solution: If it takes Allison 2 minutes to solve one easy problem, then it will take her $2 \times 5 = 10$ minutes to solve all 5 easy problems. Similarly, if it takes 4 minutes to solve a medium problem, it will take $4 \times 5 = 20$ minutes to solve all 5 medium problems, and if it takes her 7 minutes to solve one hard problem, it will take her $7 \times 5 = 35$ minutes to solve all 5 hard problems.

In total, it will take her $10 + 20 + 35 = \boxed{65 \text{ minutes}}$ to solve all 15 problems.

11. Kate is eating a square pizza. Her sister, Rachel, is eating a rectangular pizza with side lengths 4 and 9. If they both end up eating the same amount of pizza, what is the side length of the square pizza?

    Solution: First, if they ate the same amount of pizza, then the areas of their pizzas must be the same. Also, since the area of a rectangle is $length \times width$, the area of each pizza is $4 \times 9 = 36$.

    Let s be the side length of the square pizza. Since the area of a square is the side length times itself, the area of the square is $s \times s = s^2$. Then, since 36 is the area of the square pizza, $s^2 = 36$. Then, take the square root of each side of the equation to get that $s = \boxed{6}$.

12. Hannah pastes together four identical rectangles of length 4 and width 7 to form one large rectangle. What is the area of the final rectangle?

    Solution: If four identical rectangles are pasted together, the area of the large rectangle will be four times the area of the small rectangle. Then, since the area of the small rectangle is $4 \times 7 = 28$, the area of the large rectangle is $4 \times 28 = \boxed{112}$.

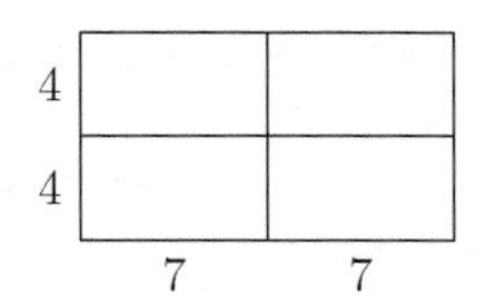

13. A school has some books. All the books can be split evenly among 6 classrooms. They can also be split evenly among 10 classrooms. What is the smallest number of books that the school can have?

    Solution: If all the books can be split evenly among 6 classrooms, the total number of books must be a multiple of 6. Similarly, the total number of books is also a multiple of 10. To find the smallest number that is a multiple of 6 and 10, list the multiples of 6 and 10 until both of them share a number.

    Multiples of 6 are: 6, 12, 18, 24, **30**, 36, ...

Multiples of 10 are: 10, 20, **30**, 40, ...

The first multiple shared by 6 and 10 is 30. Therefore, the answer is $\boxed{\text{30 books}}$.

Solution 2: If all the books can be split evenly among 6 classrooms, the total number of books must be a multiple of 6. Similarly, the total number of books is also a multiple of 10. Then, the answer to this problem is the least common multiple of 6 and 10. Since $6 = 2 \times 3$ and $10 = 2 \times 5$, the least common multiple of 6 and 10 is $2 \times 3 \times 5 = \boxed{\text{30 books}}$.

14. Ara uses a keyboard with a total of 104 keys. She cannot read yet, so she randomly presses two keys, one after another. What is the probability that both keys are letters? (There are 26 letters in the alphabet).

    Solution: First, realize that the probability of pressing a letter key on her first press is:

$$\frac{number\ of\ letter\ keys}{number\ of\ total\ keys} = \frac{26}{104} = \frac{1}{4}$$

    Similarly, the probability of pressing a letter key on her second press is also $\frac{1}{4}$. This means the odds of pressing letter keys on both presses is $\frac{1}{4} \times \frac{1}{4} = \boxed{\frac{1}{16}}$

15. A square and an equilateral triangle have side length 5. Michelle glues one edge of the equilateral triangle to one edge of the square, keeping the square and the triangle touching at only one edge. What is the perimeter of this new figure?

    Solution: First, draw a diagram similar to the one below:

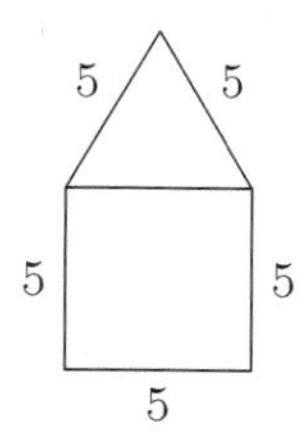

    Then, add up all of the outer sides of the figure to get the perimeter of the figure: $5+5+5+5+5=\boxed{25}$.

16. A triangle has two angles of 20° and 50°. What is the measure of the third angle?

    Solution: In order to solve this, it helps to know that the sum of the angles in a triangle is 180°. This means the third angle must be $180° - 20° - 50° = \boxed{110°}$.

17. A lake is filled with 20 fish. After every morning, 3 fish swim out of the lake and after every night, 2 fish swim in. After how many mornings will the lake first be empty?

Solution: First, start with 20 fish in the pond.

On the first morning, 3 fish swim out, so there are 17 fish remaining. On the first night, 2 fish swim in, so there are 19 fish left.

On the second morning, 3 fish swim out, so there are 16 fish remaining. On the second night, 2 fish swim in, so there are 18 fish left.

On the third morning, 3 fish swim out, so there are 15 fish remaining. On the third night, 2 fish swim in, so there are 17 fish left.

Notice how by the end of the day, the total fish count drops by one. This means that by the end of the 17th day, there will be 3 fish remaining. Then, on the 18th morning, all 3 fish will swim out, leaving no fish remaining. Therefore, it takes $\boxed{\text{18 mornings}}$ for the lake to be empty.

18. Angle A in triangle ABC has an angle measure of 42 degrees. What is the average angle measure of the other two angles of ABC?

Solution: First, it helps to know the sum of the angles in a triangle is $180°$. This means that the sum of the other two angles is $180° - 42° = 138°$. Then, to find the average of the two angles, divide the sum of the angles by 2. This means the average of the two angles is $138° \div 2 = \boxed{69°}$.

19. Two apples weigh as much as three oranges. Nine oranges weigh as much as ten pears. How many apples weigh as much as 40 pears?

Solution: In this problem, the weight ratios of the pears to oranges and oranges to apples can help convert the weight of pears to the weight of apples. Since the problem gives the number of pears, first convert the weight of the pears to the weight of oranges, then the weight of oranges to the weight of apples to get the final answer. Using this process, the result is:

$$40\ pears \times \frac{9\ oranges}{10\ pears} \times \frac{2\ apples}{3\ oranges} = \boxed{24\ apples}$$

20. $3x - 4y = 18$ and $4x - 3y = 10$. Compute $x - y$.

Solution 1: If these two equations are added together, then:

$$\begin{array}{r} 3x - 4y = 18 \\ +4x - 3y = 10 \\ \hline 7x - 7y = 28. \end{array}$$

Finally, divide both sides of the new equation by 7 to get $x - y = \boxed{4}$.

Solution 2: Another possible way to solve this problem is by calculating x and y using substitution. First, using the equation $3x-4y = 18$, find x in terms of y:

$$
\begin{aligned}
3x - 4y &= 18 \\
3x &= 4y + 18 \\
x &= \frac{4}{3}y + 6
\end{aligned}
$$

Then, $\frac{4}{3}y + 6$ can be substituted for x in the equation $4x - 3y = 10$:

$$
\begin{aligned}
4x - 3y &= 10 \\
4(\frac{4}{3}y + 6) - 3y &= 10 \\
\frac{16}{3}y + 24 - 3y &= 10 \\
\frac{7}{3}y &= -14 \\
\frac{1}{3}y &= -2 \\
y &= -6
\end{aligned}
$$

Then, since $x = \frac{4}{3}y + 6$, x must be

$$\frac{4 \times (-6)}{3} + 6 = -2$$

This means $x - y = (-2) - (-6) = \boxed{4}$

21. Two chipmunks can build a treehouse in five hours. How much time will it take for ten chipmunks to build twenty treehouses?

    Solution: If 2 chipmunks can build 1 treehouse in 5 hours, then 2×5 chipmunks (or ten chipmunks) can build 1×5 treehouses in 5 hours. To build 20 treehouses (or 5 treehouses four times), it will take the same 10 chipmunks four times as long. This means it will take them 4×5 hours, or $\boxed{\text{20 hours}}$.

22. How many squares of side length 2 can fit inside a 5 by 7 rectangle with sides parallel to the sides of the rectangle and no overlap allowed?

    Solution: In order to solve this, draw a diagram like the one shown below. (Note: each small dashed square is a 1×1 square, and each large square is a 2×2 square):

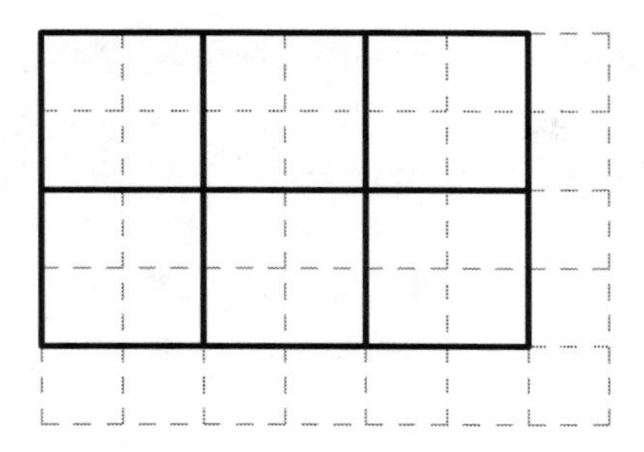

Counting the large squares in the diagram, $\boxed{6 \text{ squares}}$ with side length 2 can fit in a 5 by 7 rectangle without overlap.

23. A square frame for a square painting is made with a thickness of 1 inch. If the area of the frame is three times the area of the painting, what is the side length, in inches, of the painting?

Solution: Let p be the side length of the painting. Then, draw a diagram of the painting and frame similar to the one below:

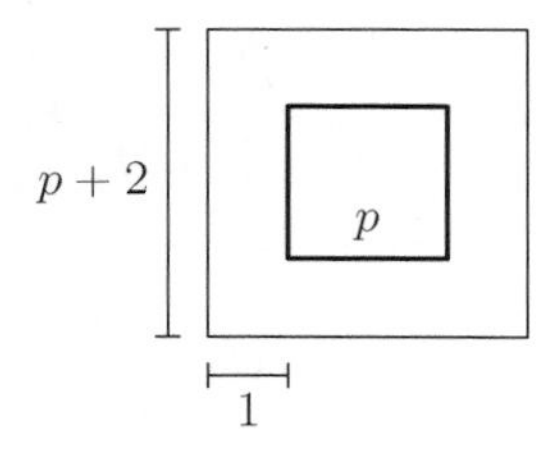

In this diagram, the painting is the inner square, and the frame is the outer square. This means the area of the painting is $p \times p = p^2$.

Then, notice that the frame adds 1 inch of width on each side of the frame, so the frame is actually a square with side length $p + 2$. This makes the area of the frame and the painting combined equal to $(p + 2)^2 = p^2 + 4p + 4$. To find the area of the frame without the painting, take the total area, and subtract the area of the painting. The area of the frame will then be $(p^2 + 4p + 4) - (p^2) = 4p + 4$.

Next, since area of the frame is three times the area of the painting, $3 \times p^2 = 4p + 4$. Subtract $4p + 4$ from each side of the equation to get $3p^2 - 4p - 4 = 0$. From here, there are two methods to solving the equation:

Option 1: Guess and Check Method.
Start with $p = 1$, and keep guessing numbers until one of them satisfies the equation $3p^2 - 4p - 4 = 0$.

If $p = 1$, then $3(1)^2 - 4(1) - 4 = 3 - 4 - 4 = -5 \neq 0$. Since -5 does not equal 0, p cannot equal 1.

Then, if $p = 2$, then $3(2)^2 - 4(2) - 4 = 12 - 8 - 4 = 0$. This equals 0, so $p = \boxed{2 \text{ inches}}$.

Option 2: Factor the Quadratic
Factoring the quadratic gives the result $3p^2 - 4p - 4 = (3p+2)(p-2) = 0$. From this, $p = -\frac{2}{3}$ or 2. Since side lengths cannot be negative, $p = \boxed{2 \text{ inches}}$.

24. Julia, Joey, and Jackie are going to Disneyland together. They each pay $158 for their ticket. When they get there, they spend $60 on lunch collectively and $72 on dinner collectively. If they split the bill equally for both meals and add their ticket price to their total, what did each person pay in total?

    Solution: First, their tickets each cost $158. Then, they split the lunch bill evenly between the 3 of them, so they each spent $\$60 \div 3 = \$20$ on lunch. And finally, if they split the dinner bill evenly between the 3 of them, they each spent $\$72 \div 3 = \$24$ on dinner. Therefore, each person spent a total of $\$158 + \$20 + \$24 = \boxed{\$202}$.

25. Isabella has a bag of 15 green, blue, and violet marbles. There are 9 blue marbles, and twice as many green marbles as violet marbles. If she randomly picks one marble, and without putting it back, randomly picks another marble, what is the probability that both marbles are green?

    Solution: First off, figure out how many marbles there are of each color. First, there are 9 blue marbles, so there are $15 - 9 = 6$ marbles that are either green or violet.

    Let g be the amount of green marbles, and let v be the amount of violet marbles. This means $g + v = 6$. However, since there are twice as many green marbles as violet marbles, then $g = 2 \times v$. Then, by replacing g with $2 \times v$ in the equation $g + v = 6$, then $2 \times v + v = 6$, or $3v = 6$. Dividing both sides of the equation by 3 results in $v = 2$, so there are 2 violet marbles.

    Then the amount of green marbles is $2 \times v$, so the total number of green marbles is $2 \times 2 = 4$.

    Then, the probability of picking a green marble on the first pull is:

$$\frac{number\ of\ green\ marbles}{number\ of\ total\ marbles} = \frac{4}{15}$$

    In the second pull, she has already taken out one marble, so there are only 14 marbles left to choose from. In addition, she has taken out one of the green marbles, so there are only 3 green marbles left. This makes the probability of pulling a green marble on the second pull $\frac{3}{14}$.

Finally, the probability of pulling a green marble both times must be the probability of picking a green marble on the first pull multiplied by the probability of picking a green marble on the second pull. This probability is $\frac{4}{15} \times \frac{3}{14} = \frac{2}{5} \times \frac{1}{7} = \boxed{\frac{2}{35}}$.

26. Jack has some two-liter water bottles and some three-liter water bottles. He wants to fill a pail of 17 liters while wasting no water. What is the least number of two-liter bottles that Jack has to use?

    Solution: First, since the number of two-liter water bottles is supposed to be as small as possible, assume zero two-liter water bottles can be used. Unfortunately, since 17 is not a multiple of 3, Jack cannot get 17 liters of water by using only 3-liter bottles.

    Then, assume one two-liter water bottle can be used. This works because five three-liter bottles gives him 15 liters of water, and one two-liter bottle gives him a total of 17 total liters of water. This makes the answer $\boxed{\text{1 two-liter bottle}}$.

27. Ilene chose a two digit number. She then reversed the digits of the number: for example, from 19 to 91. She finally added these two numbers together and got 143. What is the sum of the digits of her original number?

    Solution: First, realize that any two digit number can be expressed as $10 \times \textit{tens digit} + \textit{ones digit}$. (For example, 93 is $9 \times 10 + 3$ and 48 is $4 \times 10 + 8$.)

    Then, represent the two-digit number she chose in the form $10 \times x + y$, where x is the tens digit, and y is the ones digit. Also, since reversing the digits of the number switches the places of the ones and tens digits, the 'reverse' of this number is $10 \times y + x$.

    Then, since adding the original and reverse number together equals 143, then $(10x + y) + (10y + x) = 11x + 11y = 11 \times (x + y) = 143$.

    From here, it seems impossible to solve for x and y. However, since the problem is asking for the sum of the digits, the goal is to find the value of $x + y$. Since $11 \times (x + y) = 143$, divide both sides of the equation by 11 to get the answer: $x + y = \boxed{13}$

28. A rectangle with integer side lengths has an area of 42. How many different possible values are there for its longer side?

    Solution: First, list all possible rectangles with integer side lengths and an area of 42. This can be done by finding all of the possible divisors for 42, and pairing them up so when multiplied together, their result is 42. If done correctly, 4 rectangles are given:

    1 by 42

2 by 21
3 by 14
6 by 7

Each of the larger numbers in the pairs (42, 21, 14, and 7) can each be a possible value for the longer side of the rectangle. This means there are $\boxed{\text{4 possible values}}$ for the longer side of the rectangle.

29. Jason picks a positive integer. He multiplies his number by itself. Then, he subtracts 1. He gets the answer 399. What was Jason's original number?

    Solution: Let n be the original number he picked. If he multiplies the number by itself, the result is $n \times n = n^2$. Then, he subtracts 1 from it and gets 399, which means $n^2 - 1 = 399$. Adding 1 to each side of the equation results in $n^2 = 400$. Finally, square root each side of the equation to get $n = \pm 20$. However, n is positive, so n must equal $\boxed{20}$.

30. Mohab is made up of a circle for his head, a square for his body, and 4 congruent rectangles for his arms and legs with no overlapping regions. The circle has a radius that is equivalent to each side of the square, 4 units. The dimensions of each rectangle are 6 units by 1 unit. What is Mohab's overall area? Leave your answer in terms of π.

    Solution: First, draw a diagram representing the problem, as shown below:

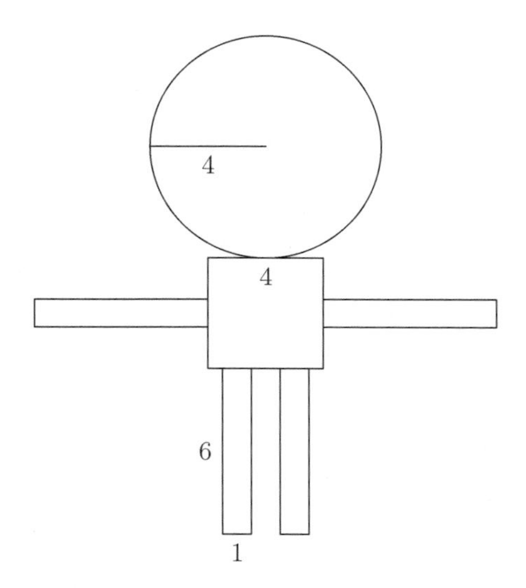

Then, calculate the area of each body part.

Since the head is a circle, the area of the circle is $\pi \times radius^2$. The radius is 4, so the area of the circle is $\pi \times 4^2 = 16\pi$.

Next, calculate the area of the body. The body is a square with a side length of 4, so the area is $4 \times 4 = 16$.

Finally, calculate the area of the limbs (arms and legs). Since the length of one limb is 6 and the width is 1, the area of one limb is $6 \times 1 = 6$. However, since there are 4 limbs, the area of all the limbs is $6 \times 4 = 24$.

Finally, adding all of the pieces together gives the final answer:

$$16\pi + 16 + 24 = \boxed{40 + 16\pi}$$

Target Round

1. How many edges does a rectangular prism have?

   Solution: The easiest way to solve this problem is to draw a rectangular prism (like the figure below), and count the number of edges. This gives a total of $\boxed{12 \text{ edges}}$.

   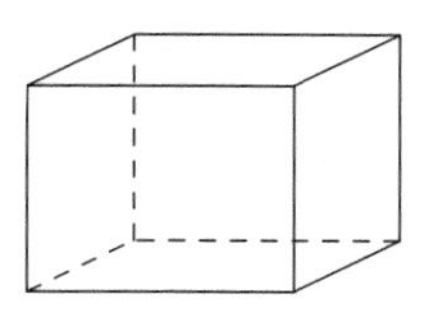

2. Jeff the sea lion is shaped awfully like a triangle. In fact, he is a triangle. Jeff is four feet wide and eight feet tall! What is half of his area?

   Solution: Since Jeff is a triangle, his area can be found by multiplying ½ times the base times the height ($Area = 1/2 \times base \times height$). This means Jeff's area is $1/2 \times 4 \times 8 = 16$. Then, half of Jeff's area is $1/2 \times 16 = \boxed{8 \text{ feet}^2}$

3. The perimeter of a regular pentagon is equivalent to the perimeter of an equilateral triangle. If the side length of the pentagon is 6, what is the side length of the triangle?

   Solution: First, calculate the perimeter of the pentagon. Since it is a regular pentagon, it has 5 sides, and each side has the same length. Since the side length of the pentagon is 6, its perimeter is $6 \times 5 = 30$.

   Then, since an equilateral triangle has 3 sides of equal length, and its perimeter must also be 30, its side length is $30 \div 3 = \boxed{10}$.

4. Bessie has a total of 7 animals who are either cows or chickens. All of her animals have a total of 22 legs. How many of Bessie's animals are cows?

   Solution: Let x be the amount of cows Bessie has, and let y be the amount of chickens she has.

   Since she has 7 cows and chickens in total, this means $x + y = 7$.

   Cows have four legs, so the total number of cow legs is $4x$. Chickens have two legs, so the total number of chicken legs is $2y$. Since her animals have a total 22 legs, $4x + 2y = 22$.

Then, take $x + y = 7$. Subtracting x from both sides of the equation results in $y = 7 - x$. Then, substitute $7 - x$ for y in the equation $4x + 2y = 22$:

$$\begin{aligned} 4x + 2y &= 22 \\ 4x + 2 \times (7 - x) &= 22 \\ 4x + 14 - 2x &= 22 \\ 2x + 14 &= 22 \\ 2x &= 8 \\ x &= \boxed{4 \text{ cows}} \end{aligned}$$

5. At Karishma's Restaurant, a meal consists of one appetizer, one main dish, and one drink. There are 5 choices for an appetizer, 10 choices for a main dish, and 9 choices for a drink. If Cathy is allergic to 2 of the appetizers and 5 of the main dishes, how many choices does she have for a meal?

   Solution: First, since she is allergic to 2 of the 5 appetizers, she only has 3 choices for an appetizer. Since Cathy is allergic to 5 of the 10 main dishes, she only has 5 choices for a main dish. She is not allergic to any of the drinks, so she has all 9 choices for drinks. Therefore the total number of choices she has for a meal is $5 \times 3 \times 9 = \boxed{135 \text{ choices}}$.

6. Jan invited her friends to her house for a party. At 9 PM, half of the people leave. At 10 PM, one-third of the remaining people leave. Finally, at 11 PM, three-fourths of the remaining people leave, leaving only Jan. How many people, including Jan, were at the party?

   Solution: The easiest way to solve this problem is to work backwards. At 11 PM, $3/4$ of the people leave, leaving only one person remaining, Jan. This means Jan is $1 - \frac{3}{4} = \frac{1}{4}$ of the people at the party at 11. Then, if 1 person is $1/4$ of the people at the party at 11 PM, there were $1 \div \frac{1}{4} = 1 \times 4 = 4$ people at the party before 11 PM.

   Then, at 10 PM, $1/3$ of the people left the party. This means $1 - \frac{1}{3} = \frac{2}{3}$ of the people stood at the party at 10 PM. Then, realize that nobody left after 10 PM and before 11 PM, so after 10 PM, there must have been 4 people at the party. Since 4 people made up $2/3$ of the party at 10 PM, there must have been $4 \div \frac{2}{3} = 4 \times \frac{3}{2} = 6$ people at the party before 10 PM.

   Finally, at 9 PM, $1/2$ of the people left the party, which means $1 - \frac{1}{2} = \frac{1}{2}$ of the people stood at the party. Also, nobody left after 9 PM and before 10 PM, so after 9 PM, there must have been 6 people at the party. Since 6 people made up $1/2$ of the party at 9 PM, there must have been $6 \div \frac{1}{2} = 6 \times 2 = \boxed{12 \text{ people}}$ at the start of the party.

7. At a certain store, 5 books and 1 pencil cost twice as much as 2 books and 2 pencils. 1 pencil usually costs 1 dollar. On a special savings day, 4 pencils are given for every purchase of a book. If Patricia needs at least 4 books and 30 pencils, what is the least amount of money she can spend?

   Solution: First, calculate the price of the book. Let the price of the book equal b. Then, if someone buys 2 books and 2 pencils, and pencils cost \$1 each, then the cost of 2 books and 2 pencils is $2b + 2$. Also, the cost of 5 books and a pencil is $5b + 1$. Then, since 5 books and 1 pencil cost twice as much as 2 books and 2 pencils:

$$\begin{aligned} 5b + 1 &= 2 \times (2b + 2) \\ 5b + 1 &= 4b + 4 \\ 5b - 4b &= 4 - 1 \\ b &= 3 \end{aligned}$$

   Then, since 4 pencils normally cost \$4, it is less expensive to pay only \$3 for a book and get 4 pencils for free on special savings day. If she buys 7 books for \$21, she also gets $7 \times 4 = 28$ free pencils. She only needs to pay another \$2 for 2 pencils to get to 30 pencils, so the least amount of money you can spend is \$21 + \$2 = $\boxed{\$23}$.

8. How many different rectangles with integer side lengths have an area of 100? Two rectangles are considered the same if one can be rotated to get the other one.

   Solution: In order to solve this problem, find the factors of 100, and pair them up to make rectangles with areas of 100. (Note: A factor of 100 is a number that can divide into 100 without leaving a remainder.) The factors of 100 are 1, 2, 4, 5, 10, 20, 25, 50, and 100. By pairing these up, $\boxed{\text{5 rectangles}}$ can be made:

   1 by 100
   2 by 50
   4 by 25
   5 by 20
   10 by 10

3rd-4th Grade 2017 Solutions

Sprint Round

---

1. To participate in the All Girls Math Competition, Casey needs to buy some pencils and erasers. A pencil costs 10 cents and an eraser costs 5 cents. If Casey wants 2 pencils and 2 erasers, how much money would Casey need, in cents?

   Solution: First, calculate the cost of the pencils. To do so, multiply the number of pencils by the price of each pencil, which is $2 \times 10 = 20$ cents. Then, calculate the cost of the erasers by multiplying the number of erasers by the price of each eraser, which is $2 \times 5 = 10$ cents. Now, add the cost of the pencils and the erasers to get the total, which is $20 + 10 = \boxed{30 \text{ cents}}$.

2. Susie went to the movies and brought twenty dollars. Her friends doubled the amount of money she had. Then, she spent \$6 on popcorn and \$3 on a soda. How much money did Susie have left?

   Solution: First, Susie's friends doubled her money, which means they multiplied it by 2. This means she has $\$20 \times 2 = \$40$. Then, she spends \$6 on popcorn and \$3 on a soda, which means she is left with $\$40 - \$6 - \$3 = \boxed{\$31}$.

3. Sally and Susie are comparing heights. Sally is 3 feet and 11 inches, while Susie is 5 feet and 1 inch. In inches, what is the difference between Sally's height and Susie's height?

   Solution: The easiest way to solve this problem is to convert both of their heights to inches. Since 1 foot is 12 inches, 3 feet must equal $83 \times 12 = 36$ inches, so Sally's height is $36 + 11 = 47$ inches. Also, 5 feet is $5 \times 12 = 60$ inches, so Susie's height is $60 + 1 = 61$ inches.

   Finally, subtracting their heights gives the difference between their heights, which is $61 - 47 = \boxed{14 \text{ inches}}$.

4. Jenna was making a birthday cake, so she buys 2 dozen eggs. If she had 5 eggs left over, how many eggs did she use to make the cake?

   Solution: First, realize that a dozen is equal to 12. This means she bought $2 \times 12 = 24$ eggs. Since there are 5 eggs left over, Jenna must have used $24 - 5 = \boxed{19 \text{ eggs}}$.

5. Tyler gives Emily a \$250 Verizon gift card. If Verizon charges a \$50 one-time fee to get a phone number, and then charges \$40 per month for a phone plan, how many full months can the gift card buy if Emily does not have a phone number?

Solution: First, since Emily does not have a phone number, she must pay \$50 for a phone number, leaving her with \$250 − \$50 = \$200. Then, since it costs \$40 per month for the plan, the gift card will last her $\$200 \div 40\ \frac{dollars}{month} = \boxed{5 \text{ months}}$.

6. Cheryl receives a KFC coupon in the mail. The coupon gives 50% off a bucket of fried chicken. If a bucket of fried chicken normally costs \$5.00, how much money will the bucket of chicken cost with the coupon?

   Solution: If the coupon is 50% off the normal price, this means the bucket of fried chicken is $100\% - 50\% = 50\%$ of the original price. Then, note that $50\% = \frac{50}{100} = \frac{1}{2}$, so the fried chicken bucket is half its original price, which is $\$5.00 \times \frac{1}{2} = \boxed{\$2.50}$.

7. If Iris rolls a six-sided die, what is the probability that she rolls a six? (Please answer using a fraction.)

   Solution: In order to write probability as a fraction, the number of ways to get the desired outcome goes in the numerator and the total number of possible outcomes goes in the denominator. There is only one way to roll a 6 on a die, so the numerator is 1. There are 6 possible numbers to roll on a die, so the denominator is 6. This makes the probability of rolling a 6 equal to $\boxed{\frac{1}{6}}$.

8. Jackie has a bunch of identical square Lego pieces, so she makes them into a rectangle that is 6 pieces wide and 8 pieces long. Then, using the same Lego pieces, she made a rectangle that was 4 pieces wide. How many pieces long is the new rectangle?

   Solution: First, realize that the area of a rectangle is equal to its length times its width. ($Area = length \times width$) This means Jackie has $6 \times 8 = 48$ Lego pieces. Then, since the area of the new rectangle must be the same, this means the new rectangle is also made of 48 Lego pieces. Finally, since the rectangle is 4 pieces wide, the length must be $48 \div 4 = \boxed{12 \text{ Lego pieces}}$.

9. Prom King and Queen are being decided at the All Girls Math Tournament Annual Dance. Jane receives 5 more votes than Karen's amount of votes tripled. If Jane received 23 votes, then how many votes did Karen receive?

   Solution: To set up the equation, let k be equal to the number of votes Karen received. If Karen's votes are tripled (or multiplied by 3), and 5 more votes are added, then that equals the number of votes Jane received. This means Jane's votes are equal to $3k + 5$.

   Then, since Jane got 23 votes, $3k + 5 = 23$. Then, subtract 5 from both sides of the equation to get $3k = 18$. Finally, divide both sides by 3 to get $k = \boxed{6 \text{ votes}}$.

10. For the All Girls Tournament, Drew has to write 30 problems for the Sprint Round and 8 problems for the Target Round. If it takes 5 minutes to write a Sprint Round Problem, and 10 minutes to write a Target Round Problem, how many minutes does it take for Drew to finish writing all of his problems?

    Solution: First, the amount it will take to write the Sprint Round is the amount of problems on the Sprint Round multiplied by the time it takes to write one Sprint Round problem, which is $30 \times 5 = 150$ minutes.

    Then, the amount of time it takes Drew to write the Target Round is the amount of problems on the Target Round multiplied by the time it takes to write one Target Round problem, which is $8 \times 10 = 80$ minutes.

    Finally, the total time it took him to finish writing all the problems is the amount of time it took to write the Sprint and Target Round combined, which is $150 + 80 = \boxed{\text{230 minutes}}$.

11. One day, Ms. Chou decided to measure the height of her nine students. Their heights are 49 in, 52 in, 52 in, 52 in, 53 in, 53 in, 54 in, 54 in, and 56 in. What is the positive difference between the median height and the mode height, in inches?

    Solution: The mode height is the height that is the most common height. Since 52 inches is the height of 3 students, and no other height appears more frequently, 52 inches is the mode.

    Then, to find the median, list all the heights in numerical order, and the number in the middle of the list is the median. Since 53 inches is in the middle of the list, it is the median.

    The difference between the median and mode is $53 - 52 = \boxed{\text{1 inch}}$.

12. How many lines of symmetry does the following figure have?

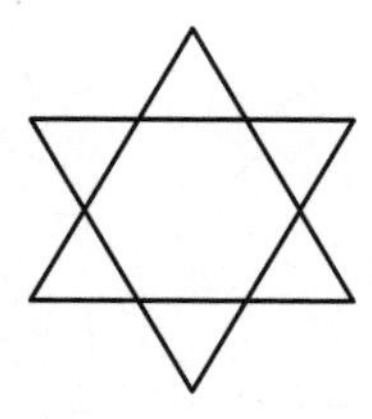

    Solution: The figure has $\boxed{\text{6 lines of symmetry}}$, as demonstrated below:

13. Kourtney goes shopping at Justice. A shirt there costs \$10, and a skirt costs \$12. If Kourtney has exactly enough money to buy 6 shirts, how many skirts could she buy?

    Solution: If she has just enough money to buy 6 shirts, she has $\$10 \times 6 = \$60$. Since a skirt costs \$12, she has enough to buy $\$60 \div \$12 = \boxed{5 \text{ skirts}}$.

14. Nessa wants to tile a room that is 5 feet long and 5 feet wide. If the tiles are each 15 inches long and 15 inches wide, how many tiles does she need to tile the floor of the room?

    Solution: Since a foot is equal to 12 inches, the room is $12 \times 5 = 60$ inches long and 60 inches wide. This means that Nessa can fit $60 \div 15 = 4$ tiles along the length, and $60 \div 15 = 4$ tiles along the width. This means she needs $4 \times 4 = \boxed{16 \text{ tiles}}$ to cover the entire floor.

15. If $3 \# 1 = \frac{3*1}{3-1}$, then what is $6 \# 4$?

    Solution: To find $6 \# 4$, substitute $6 \# 4$ for $\frac{6*4}{6-4}$. Then, simplify this expression to get the final answer:

    $$\frac{6*4}{6-4} = \frac{24}{2} = \boxed{12}$$

16. A cosmetic product, called Barbie's Amazing Lip Gloss, either works or does not work. Chloe buys 20 of the lip gloss containers, and she soon discovers that 16 containers work but 4 do not. If the 20 lip gloss containers are put in a bag, and Chloe chooses one out of the bag at random, what is the probability she grabs a lip gloss that does NOT work? (Please answer using a fraction in lowest terms.)

    Solution: Recall that in order to express probability as a fraction, the number of ways to fulfill the condition goes in the numerator, and the total number of possible outcomes goes in the denominator.

    There are 4 lip glosses that do not work, so there are 4 ways for her to pick a non-working lip gloss. There are 20 lip glosses, so there are

20 total possibilities. This means the probability of picking the lip gloss is $\frac{4}{20}$, which simplifies to $\boxed{\frac{1}{5}}$.

17. All of the small triangles in the figure below are congruent. If one of the triangles has an area of 4, what is the area of the entire figure?

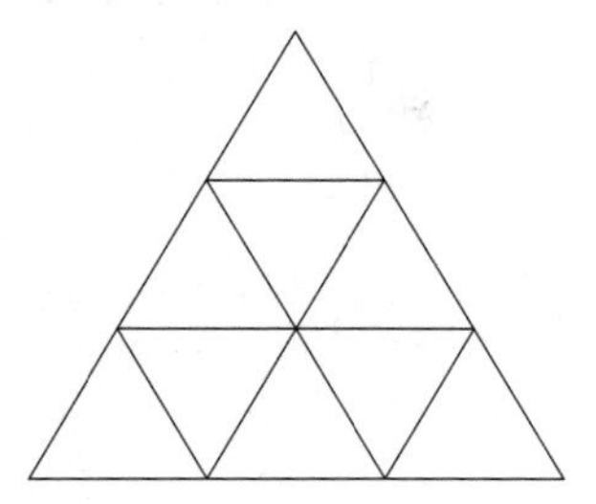

Solution: Since all of the small triangles are congruent, they all have the same area. So, to calculate the area of the big triangle, multiply the area of each small triangle by the total number of small triangles. This results in $4 \times 9 = \boxed{36}$.

18. In 2017, April Fool's is on a Saturday. In 2018, what day of the week will April Fools be on? (There are 365 days in a year.)

Solution: Divide 365 by 7 to find how many whole weeks have passed, and the remainder will be how many "extra" days are left. Since $365 \div 7 = 52$ with a remainder of 1, there is one extra day. To add this extra day, move 1 day forward from Saturday, which is $\boxed{\text{Sunday}}$.

19. Mrs. Runge has a certain number of desks in her classroom. If she arranges the desks into rows of 3, she will have 2 desks that are not in a row. If she arranges the desks into rows of 5, she will have 4 desks that are not in a row. If she has less than 25 desks, how many desks does she have?

Solution: Let n be the number of desks Mrs. Runge has. If Mrs. Runge had one more desk, then she could arrange the desks into rows of three with no desks left over, and into rows of five with no desks left over. This means $n+1$ is a multiple of 3 and 5. The only number that is a multiple of 3 and 5 and is less than $25+1$ is $3 \times 5 = 15$. Then, $n+1 = 15$, which means $n = \boxed{14 \text{ desks}}$.

20. What is the ones digit of 5×372?

Solution: First, realize that the ones digit is the last digit in a whole number, or the digit before the decimal point. There are two ways to solve this problem. The long way is to multiply 5 by 372. The more efficient way is to realize that $5 \times 372 = 5 \times 370 + 5 \times 2$. However, since the ones digit of 5×370 is zero, it will not affect the ones digit of the final answer.

This means the ones digit of 5×2 is the same as 5×372. Finally, the ones digit of $5 \times 2 = 10$ is 0, so the answer is $\boxed{0}$.

21. A certain rectangle is 4 inches long and 3 inches wide. If the length of the rectangle is doubled, and the width of the rectangle is tripled, what is the area of the new rectangle?

    Solution: If the length of the rectangle is doubled, the new length will be $4 \times 2 = 8$ inches. Also, if the width is tripled, the new width is $3 \times 3 = 9$ inches. Now, the new rectangle's area can be easily found by multiplying the new width and the new height, which is $8 \times 9 = \boxed{72 \text{ in}^2}$.

22. Joe has a house of cards that is built with cards stacked like the figure below. If he makes a card house that has 4 stories, how many cards does he need to put in it?

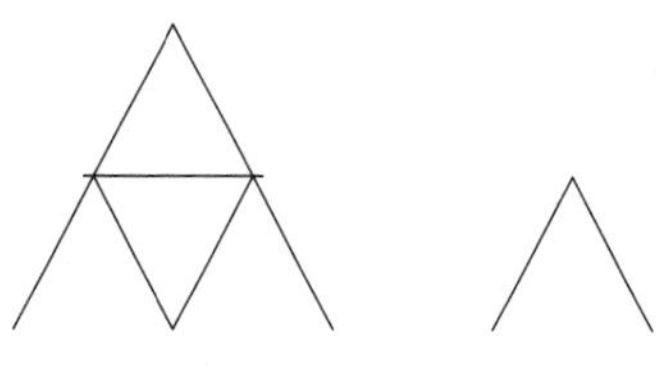

2 story card house 1 story card house

Solution 1: The easiest way to solve this problem is to draw the 4-story card house and count the cards. As demonstrated by the figure below, there are 2 cards in the top story, 4 cards in the 2nd story, 6 cards in the 3rd story, and 8 cards in the 4th story. There is also 1 card between the 1st and 2nd stories, 2 cards between the 2nd and 3rd stories, 3 cards between the 3rd and 4th stories. This means the total amount of cards is $2+4+6+8+1+2+3 = 20+6 = \boxed{26 \text{ cards}}$.

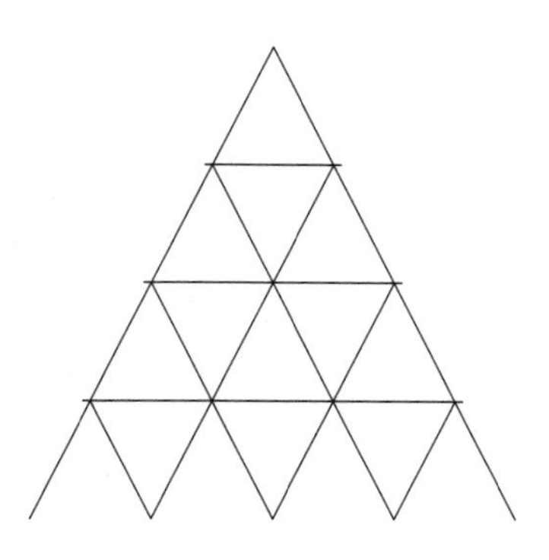

Solution 2: In the figure above, there are 2 cards in the top story. In the 2nd story, there are 2 pairs of 2 cards, and one extra card above the two pairs, making a total of 5 cards. In the 3rd story, there are 3 pairs of 2 cards, and two extra cards above the two pairs, making a total of 8 cards.

The number of cards in these three stories create the pattern where the difference in cards between each story is 3 cards. Therefore, there are $8 + 3 = 11$ cards in the 4th story, so there are $2 + 5 + 8 + 11 = \boxed{26 \text{ cards}}$ in the 4 story card house.

23. Billy Jean draws 5 squares side by side, as shown in the figure below. If the squares have side lengths of 1, 2, 3, 4, and 5, in that order, what is the area of the figure?

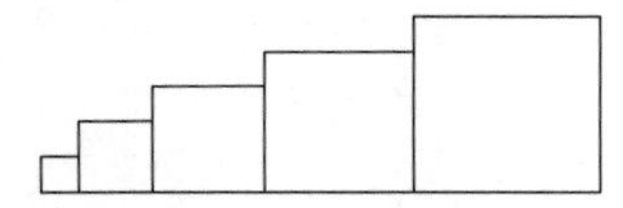

Solution: The area of the figure is the sum of the area of all 5 squares. To calculate the area of a square, multiply the side length of the square by itself. ($Area = side \times side$)

This means the area of the entire figure is equal to:

$$\begin{aligned}(1 \times 1) + (2 \times 2) + (3 \times 3) + (4 \times 4) + (5 \times 5) \\ = 1 + 4 + 9 + 16 + 25 \\ = \boxed{55}\end{aligned}$$

24. What is the perimeter of Billy Jean's figure in the previous problem?

Solution: First, realize that the bottom side is equal to $1 + 2 + 3 + 4 + 5 = 15$, and the right side is equal to 5. That leaves the 5 tops of each squares, and the bold lines in the figure below. However, the sum of the tops of the square is also the sum of the side lengths of the squares, or $1 + 2 + 3 + 4 + 5 = 15$. Also, realize that the length of a bold line is the side length of a larger square minus the length of the square smaller than it. This means every bold line has a length of 1. This means the total perimeter is $15 + 5 + 15 + (1 \times 5) = \boxed{40}$.

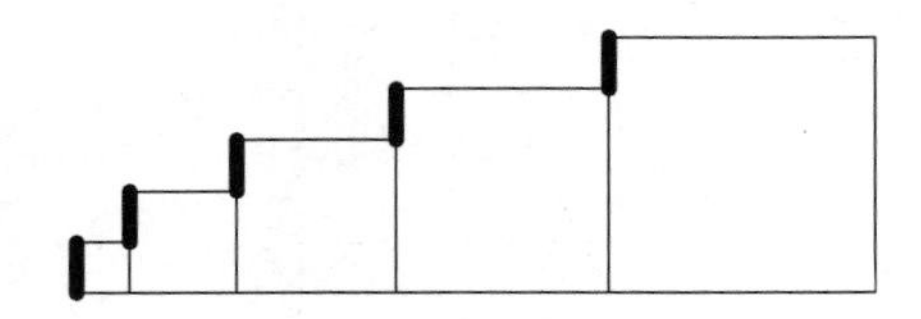

25. Lara and Natasha had 120 marbles in total. After Lara gave Natasha 30 marbles, Natasha had twice as many marbles as Lara. How many marbles did Natasha have *before the exchange*?

    Solution: Let x be equal to Lara's marbles after the exchange. Since Natasha has twice as many marbles as Lara after the exchange, she must have $2x$ marbles.

    Since the sum of their marbles is equal to 120, then $x+2x = 3x = 120$. Then, dividing both sides of the equation by 3 gives that $x = 40$ marbles.

    Since Natasha had $2x$ marbles, she has $2 \times 40 = 80$ marbles after the exchange. Since Lara gave her 30 marbles during the exchange, she only had $80 - 30 = \boxed{50 \text{ marbles}}$ before the exchange.

26. Carla is planning to use her allowance of $20 every week in order to buy a new phone, which costs $200. However, the price of the phone rises by 50%, but luckily she gets a 20% discount on the new price. How many weeks will it take her to save up enough money to buy the phone?

    Solution: First, the price of the phone raises by 50%, which means that the new price is $200 \times 50\% = 200 \times \frac{50}{100} = 200 \times \frac{1}{2} = 100$ dollars more expensive. This makes the phone's new price $200 + 100 = 300$ dollars.

    However, if she gets a 20% discount, she has to pay $100\% - 20\% = 80\%$ of the new price. This means she will have to pay $300 \times 80\% = 300 \times \frac{80}{100} = 300 \times \frac{4}{5} = 240$ dollars for the phone. Finally, if she makes $20 a week, she will have to save up for $\$240 \div 20 \ \frac{dollars}{week} = \boxed{12 \text{ weeks}}$.

27. If a plap is three plups, and a plup is two ploops, how many ploops does it take to make three plaps?

    Solution: In this problem, use the ratios of the plaps, plups, and ploops to convert from one to another. In each fraction that is multiplied, the denominator cancels out the numerator of the fraction before it, therefore converting from one unit to another. Since the amount of plaps is given, first convert plaps to plups, and then plups to ploops to get our final answer. This process creates the equation:

$$3\ plaps \times \frac{3\ plups}{1\ plap} \times \frac{2\ ploops}{1\ plup} = \boxed{18\ ploops}$$

28. Jenna has a square that is six inches long on each side. She also has 4 triangles that have six-inch bases, and are 4 inches tall. If she glues one triangle to each side of the square, what is the area of the figure?

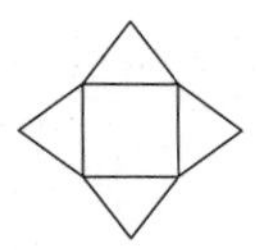

Solution: The area of the figure is the area of the square plus the area of the four triangles. To calculate the area of a square, multiply the side length of the square by itself. ($Area = side \times side$) Since the square has a side length of 6, the area of the square is $6 \times 6 = 36$.

Then, since the area of a triangle is equal to $\frac{1}{2} \times base \times height$, the area of one triangle is $\frac{1}{2} \times 6 \times 4 = 12$. This means the area of all four triangles is $12 \times 4 = 48$.

Finally, the total area of the figure is the sum of the area of the triangles and squares, which is $36 + 48 = \boxed{84 \text{ in}^2}$.

29. If four people working for four hours can complete eight jobs, then how many jobs can eight people complete in eight hours?

    Solution: If 4 people working for 4 hours can complete 8 jobs, then $4 \times 2 = 8$ people working for 4 hours can complete $8 \times 2 = 16$ jobs. If 8 people working for 4 hours can complete 16 jobs, then 8 people working for $4 \times 2 = 8$ hours can complete $16 \times 2 = \boxed{32 \text{ jobs}}$.

30. Miki is doing laps around her school's football field, which is 320 feet long and 160 feet wide. If it takes her 20 seconds to run the width of the football field, how many seconds does it take for her to run a complete lap around her school's football field?

    Solution: First, calculate the perimeter of the football by using this formula: $2 \times (length + width) = Perimeter$. If the length is 320 feet, and the width is 160 feet, the perimeter is $2 \times (320 + 160) = 960$ feet. Then, if she can run the width of the football field in 20 seconds, she runs at a speed of $160\ feet \div 20\ seconds = 8\frac{feet}{second}$. If she runs at a speed of $8\frac{feet}{second}$, it will take her $960\ feet \div 8\frac{feet}{second} = \boxed{120 \text{ seconds}}$ to run a complete lap around the football field.

Target Round

1. Kevin really loves reading the Magic Tree House series, and he reads it every day. On weekdays, he reads 25 pages a day, and on weekends, he reads 45 pages a day. If a Magic Treehouse book is 150 pages long, how many complete books can Kevin finish in five weeks?

   Solution: First, calculate how many pages Kevin reads in a week. Note that there are 5 weekdays (Monday-Friday) and 2 days in the weekend (Saturday and Sunday). During weekdays, he reads 25 pages a day, so he reads $25 \times 5 = 125$ pages in a week's worth of weekdays. Also, since he reads 45 pages per day over the weekend, he reads $45 \times 2 = 90$ pages on weekends. This means Kevin can read $125 + 90 = 215$ pages in a week.

   If he reads 215 pages in one week, then Kevin will read $215 \times 5 = 1075$ pages in five weeks. Since each Magic Treehouse book is 150 pages, he can read $1075 \div 150 = 7$ books with 25 pages left over. However, since the question asks for the number of complete books Kevin can read, the answer is $\boxed{\text{7 books}}$.

2. How many triangles are in the figure below? (Hint: There are more than five!)

   Solution: Label each intersection with a point, as shown below. Then, the following triangles can be made: $\triangle AFG, \triangle BGH, \triangle CHI, \triangle DIJ, \triangle EFJ, \triangle ACJ, \triangle BDF, \triangle CEG, \triangle DAH$, and $\triangle EBI$. Therefore, there are a total of $\boxed{\text{10 triangles}}$.

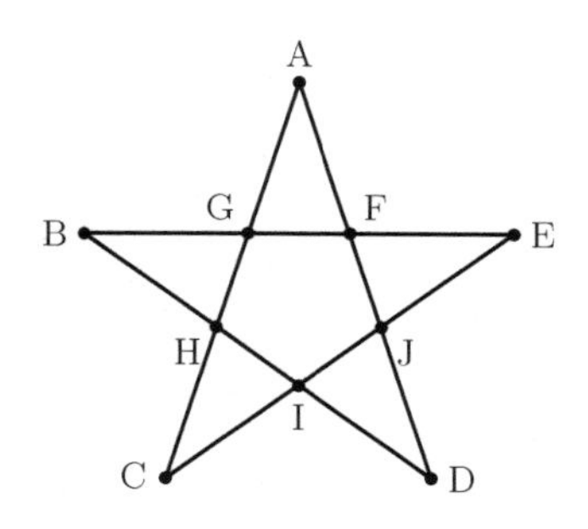

3. Chyna wanted to help with her school's bake sale, so she decided to make some cookies. In Chyna's cookie recipe, she needs $1\frac{1}{2}$ cups of flour to make 12 cookies. If she wants to make 72 cookies, how many cups of flour does she need?

   Solution: Since the original recipe makes 12 cookies and Chyna wants to make 72 cookies, the recipe must be multiplied by $72 \div 12 = 6$. Therefore, she must use $1\frac{1}{2} \times 6 = \frac{3}{2} \times 6 = \frac{18}{2} = \boxed{\text{9 cups}}$ of flour to make 72 cookies.

4. There are about 7 billion people in the world, and about $\frac{3}{5}$ of them live in Asia. If $\frac{2}{3}$ of Asians do NOT live in China, about how many people live in China?

   Solution: First, $\frac{3}{5} \times 7,000,000,000 = 4,200,000,000$ people live in Asia. If $\frac{2}{3}$ of Asians do NOT live in China, then $1 - \frac{2}{3} = \frac{1}{3}$ of Asians do live in China. This means $\frac{1}{3} \times 4,200,000,000 = \boxed{1{,}400{,}000{,}000}$ or $\boxed{\text{1.4 billion}}$ people live in China.

5. At Math Tourney Prep School, there are 150 elementary and middle school students. If 40% of students are boys, $\frac{2}{3}$ of students are in middle school, and 30% of middle school students are boys, how many students are elementary school girls?

   Solution: First, since $40\% = \frac{40}{100} = \frac{2}{5}$ of the students are boys, then $\frac{2}{5} \times 150 = 60$ students at the school are boys, and $150 - 60 = 90$ students are girls. Also, since $\frac{2}{3}$ of the students are in middle school, then $\frac{2}{3} \times 150 = 100$ of the students are in middle school.

   Next, since $30\% = \frac{30}{100} = \frac{3}{10}$ of the students *in the middle school* are boys, then $\frac{3}{10} \times 100 = 30$ students in the middle school are boys. This means the remaining $100 - 30 = 70$ students in the middle school are girls. Finally, since there are 90 girls in the school, and 70 of them are in middle school, $90 - 70 = \boxed{\text{20 students}}$ are elementary school girls.

6. Since the number four is an unlucky number in China, buildings don't use the number four when they are numbering floors. For example, they don't have a floor labeled as "14" because 14 has a four in it. If the top floor of a building in China is labeled "50," how many floors does the building actually have?

   Solution: The easiest way to do this is to count the number of floors with a 4 in it, and subtract that number from the 50 possible floors. Also, when counting numbers with a "4", it is important to realize that numbers with a 4 in the tens place, like "42", have a 4 in them and should not be counted in the floor total.

   Therefore, the numbers that do not have a floor are 4, 14, 24, 34, and 40-49. This gives a total of 14 numbers that do not have a floor.

Finally, if there are 50 numbers, and 14 of them do not have a floor, then the building is $50 - 14 = \boxed{36 \text{ floors}}$ tall.

7. Madilyn is playing a basketball game with her friends. In their basketball game, there are no free throws, so Madilyn can only make two-point shots or three-pointers. If she makes 12 shots, and scores 28 points, how many three-pointers did she make?

   Solution: Let x be the number of two-point shots Madilyn makes, and let y be the number of three-pointers she makes.

   Then, if she makes a total of 12 shots, then $x + y = 12$. Also, since she scores a total of 28 points, then 2 times the number of two point shots made plus three times the number of three-pointers made must equal 28, or $2x + 3y = 28$.

   Then, take $x + y = 12$. Subtracting y from both sides of the equation results in $x = 12 - y$. Then, substitute $12 - y$ for x in the equation $2x + 3y = 28$:

$$\begin{aligned} 2x + 3y &= 28 \\ 2 \times (12 - y) + 3y &= 28 \\ 24 - 2y + 3y &= 28 \\ y + 24 &= 28 \\ y &= \boxed{4 \text{ three-pointers}} \end{aligned}$$

8. Matt and Linna were hungry one day, so they decided to have a chip eating-contest. Since Matt decides to be kind, he gives Linna a 60-second head start. Linna eats at a rate of 6 chips per minute, and Matt eats at a rate of 8 chips per minute. How many chips will Matt have eaten when he catches up to Linna?

   Solution: First, since Linna gets a 30-second head start, or ½ minute start, she eats $1/2 \times 6 = 3$ chips when Matt starts eating. Then, since Matt eats 8 chips per minute, he eats $8 - 6 = 2$ more chips than Linna does each minute. Then, since Linna starts 3 chips ahead, and Matt catches up at a rate of 2 chips per minute, it will take $\frac{3}{2}$ minutes for Matt to catch up with Linna.

   Finally, since Matt eats 8 chips per minute and has eaten chips for $\frac{3}{2}$ minutes, he must have eaten $8 \times \frac{3}{2} = \boxed{12 \text{ chips}}$ by the time he catches up to Linna.

5th-6th Grade 2014 Solutions

Sprint Round

---

1. A certain poodle usually drinks water at a rate of 1 liter per 7 minutes. However, since the poodle is very thirsty today, its drinking rate is tripled. If we leave six-sevenths of a liter out for the poodle, in how many minutes will it finish?

   Solution: The normal rate the poodle drinks water at is 1 liter every 7 minutes, or $\frac{1}{7}$ liters per minute. However, since the rate triples, it drinks at three times the normal rate, or $3 \times \frac{1}{7} = \frac{3}{7}$ liters per minute. Since the bowl has $\frac{6}{7}$ liters of water, and the poodle drinks $\frac{3}{7}$ liters every minute, it will take the poodle $\frac{6}{7}\ liters \div \frac{3}{7}\frac{liters}{min} = \boxed{2 \text{ minutes}}$ to drink all the water.

2. Olivia likes to play basketball every other day of the week. If she starts playing on a Tuesday, how many times will she play basketball in two weeks, starting Tuesday?

   Solution: Since there are seven days in one week, there are 14 days in two weeks. If Olivia only plays basketball every other day, she plays basketball half the time. If she plays basketball one-half of the time in 14 days, she plays basketball $\frac{1}{2} \times 14 = \boxed{7 \text{ times}}$ in two weeks.

3. If x is the greatest prime factor of 36 and y is the greatest prime factor of 27 then what is $x + y$?

   Solution: First, factorize 27 and 36 into the only prime factors. To prime factorize, express these numbers using only the multiplication of prime numbers. The factorization of 27 is $3 \times 3 \times 3$, so the greatest prime factor of 27 is 3. The factorization of 36 is $2 \times 2 \times 3 \times 3$, so the greatest prime factor of 36 is 3. Adding these prime factors gives the answer, which is $3 + 3 = \boxed{6}$.

4. Two concentric circles are drawn such that one circle's radius is 3 times the other. If the area between the two circles is 72π, what is the radius of the larger circle? (Two circles are concentric if they share the same center.)

   Solution: First, realize that the area between the circles (which is the area shaded below) can be found by subtracting the two areas of the two concentric circles. If the inner radius is r, the outer radius is three times that, or $3r$. The area between the circles is then the area of the larger circle minus the area of the smaller circle, or:

$$(3r)^2\pi - r^2\pi = 9r^2\pi - r^2\pi = 8r^2\pi$$

This means that $8r^2\pi = 72\pi$. Dividing both sides of the equation by 8π results in $r^2 = 9$. Then, square root both sides to get $r = 3$. Finally, calculate the length of the larger radius by multiplying the smaller radius by 3: $3 \times 3 = \boxed{9}$.

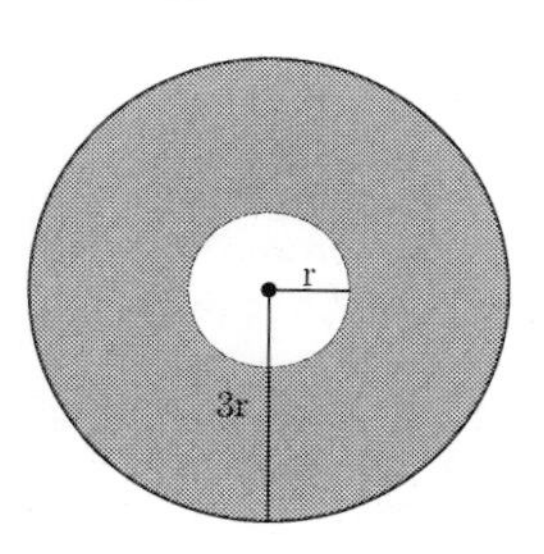

5. When a turtle wants to cross a river, he has to pay \$4. If he crosses the river 5 times on Monday, how much did he pay on Monday?

   Solution: If he crosses five times, and he has to pay \$4 each time he crosses, he pays a total of $\$4 \times 5 = \boxed{\$20}$.

6. Serge has a crush on Pam and decides to buy her 5 yards of yarn (and some roses). However, the yarn can only be bought in inches. How many inches of yarn does he need to buy?

   Solution: First, find how many inches are in one yard. Since there are 3 feet in a yard, and 12 inches in one foot, there are $12 \times 3 = 36$ inches in a yard. This means if Serge needs to buy 5 yards of yarn, he is buying $36 \times 5 = \boxed{180 \text{ inches}}$ of yarn for Pam.

7. If a bottle of sunscreen costs \$8 and Sue has a coupon that gives a 25% discount, how many dollars do she pay if she uses the coupon?

   Solution: First, calculate the discount by finding 25% of \$8.00. $25\% = \frac{25}{100} = \frac{1}{4}$, so the coupon saves her $\frac{1}{4} \times \$8 = \$2$. This means the sunscreen costs $\$8 - \$2 = \boxed{\$6}$ if she uses the coupon.

8. What is the smallest positive integer with four factors?

   Solution: An easy way to approach this problem is to count up from 1, and find the first number with four factors.

   1 has one factor (which is itself).

   2 and 3 are prime, so they only have 2 factors.

   4 has 3 factors: 1, 2, and 4.

   5 is prime, so it only has two factors.

   Then, 6 has 4 factors (1, 2, 3, and 6), so the smallest positive integer with 4 factors is $\boxed{6}$.

Solution 2: There are two ways for a number to have four factors: if it is the product of two primes, or if it is a cube of a prime. This solution tests and explains both of these cases below:

Allow p and q to be distinct prime factors, $p \times q$ has 4 positive factors: 1, p, q, $p \times q$. The smallest two primes are 2 and 3, so the smallest number in the form $p \times q$ is $2 \times 3 = 6$.

Allow n to be a prime factor, n^3 has 4 positive factors: 1, n, n^2, and n^3.The smallest prime number is 2, so the smallest number in the form n^3 is $2^3 = 8$.

Since $6 < 8$, the smallest positive integer with four factors is $\boxed{6}$.

9. Allison took a series of tests during her freshman year. Her test scores were 92, 85, 87 and 100. However, instead of 100, her test should have been a 98.5. What is the new median? Express your answer as a decimal rounded to the nearest tenth.

   Solution: To calculate the median, list all the numbers in numerical order, and the number in the middle of the list is the median. However, if there are an even amount of numbers, the median is the average of the two middle numbers.

   First, start by listing her accurate tests scores in numerical order, like so: 85, 87, 92, 98.5. Since there are an even amount of numbers, the median is the average of the two middle numbers, which are 87 and 92. The average of 87 and 92 is $(87 + 92) \div 2 = \boxed{89.5}$.

10. I am a number between 1 and 100. You get the same number if you multiply me by 4 or add 198 to me. What number am I?

    Solution: First, assume the number is "x". Then, multiplying the number by 4 is equal to $4x$, and adding 198 to the number is equal to $x + 198$. Since $4x$ is the same value as $x + 198$, $4x = x + 198$. If x is subtracted from each side, the result is $3x = 198$. Then, divide 3 from each side of the equation to get $x = \boxed{66}$.

11. The width of a rectangular prism is 3. Its length and height are both 5. What is its surface area?

    Solution: Surface area is defined as the area of each face of the prism added together. There are 6 sides of a rectangular prism. 2 sides have dimensions 5×5, so they have an area of $5 \times 5 = 25$. The other 4 sides have dimensions 5×3, and an area of $5 \times 3 = 15$. Then, the total surface area is then $25 \times 2 + 15 \times 4 = 50 + 60 = \boxed{110}$.

12. A dog runs toward a ball 30 feet away. If the dog is halfway there, how much farther does he have to run to get the ball and get back to his starting point (in feet)?

Solution: First, find how much further the dog has to run to make it to the ball. If the dog is halfway there, and the ball is 30 feet away, that means the dog has moved $30 \div 2 = 15$ feet. This also means that the dog still has 15 feet to go until he reaches the ball.

Next, once the dog reaches the ball, he must return back to his original position. Since the ball was 30 feet away from his starting point, he has to run 30 feet to get back to where he started.

Lastly, add the 15 meters it took to reach the ball to the 30 meters it took to get back to the starting point: $15 + 30 = \boxed{45 \text{ feet}}$.

13. If the beaver slaps his tail on water 4 times in 5 seconds, then pauses for 5 seconds, then repeats the cycle, how many times does he slap his tail on the water in one minute?

    Solution: If the beaver slaps the water 4 times in five seconds, and then doesn't slap it for five more seconds, the beaver is slapping the water 4 times every 10 seconds.

    Next, find how many times this cycle happens in a minute. Since 1 minute is 60 seconds, the beaver repeats this cycle $60 \div 10 = 6$ times in a minute. Finally, if one cycle has 4 slaps, the beaver must slap the water a total of $4 \times 6 = \boxed{24 \text{ times}}$ in a minute.

14. If the lengths of two of the sides of an isosceles triangle are 6 and 12, then what is the perimeter of this triangle?

    Solution: First, define an isosceles triangle. An isosceles triangle has two sides of equal length.

    In this problem, there are two options. The first option is that there are two sides of the triangle that have lengths of 6. The second option is that there are two sides which both have lengths of 12.

    However, in the first option, if there is one side which is 12, then the two sides which are 6 would form a straight line next to the line with length 12 because $6 + 6 = 12$.

    Therefore, the triangle must have two sides which are 12 and one side which is 6. Now, add up all the sides to find the perimeter: $12 + 12 + 6 = 30$. Therefore, the triangle has a perimeter of $\boxed{30}$.

15. If there are 300 cubbyholes and Joe has 301 flyers, what is the probability that one cubby will have at least two flyers?

    Solution: If each cubbyhole is filled with one flyer, than that will leave $301 - 300 = 1$ flyer left over. However, since all cubbies are occupied, putting the extra flyer in another cubby will cause at least one cubby to have two flyers. This means that the probability that one cubby will have at least two flyers is $\boxed{100\% \text{ or } 1}$.

16. The length and width of a rectangle are both doubled. What is the percent increase between the new area and the original area?

    Solution: First, let the length of the rectangle be l and let the width be w. Then, remember that the area of the rectangle is its length times its width. $(A = l \times w)$ If the rectangle's length and width are doubled, its new area is $(2 \times l) \times (2 \times w) = 4 \times l \times w = 4 \times A$. This means the new rectangle is 4 times as large as the old one.

    To calculate the percent increase, find the difference between the new area and the original area, and divide it by the original area. Since the new rectangle is 4 times the area as the original rectangle, the percent increase equals:

$$(4 \times A - A)/A = 3A/A = 3 = \boxed{300\%}$$

17. A spinner has 24 sectors, numbered 1, 2, 3, ... 24. What is the probability that, after spinning the spinner once, it lands on either a prime number or an odd number? Express your answer as a common fraction.

    Solution: All primes greater than 2 are odd, so the sectors it should land on are all odds and 2. The number of odds are ½ the total number of sectors, or $1/2 \times 24 = 12$. Then, add 1 extra sector to 12 (because of sector 2) to get a total of 13 possible sectors. Then, divide that over the total number of sectors to get a probability of $\boxed{\frac{13}{24}}$.

18. In a total of 15 numbers, the median is 38. The largest number is 93. If the largest number is changed to 97, what is the median?

    Solution: To calculate the median, list all the heights in numerical order, and the number in the middle of the list is the median. However, if there are an even amount of numbers in a set, the median is the average of the two middle numbers.

    In a set of fifteen numbers, the median is the middle number, or the 8th number. The largest number being changed to an even larger number will not affect the order of the first 14 numbers, so the new median remains at $\boxed{38}$.

19. The product of all 4 sides of a square is 1296. What is the sum of all four of the square's sides?

    Solution: Let the side length of the square be s. Then, $1296 = s \times s \times s \times s = s^4$. To find what s is, it helps to find the prime factorization of 1296:

$$\begin{aligned} 1296 &= 2 * 648 \\ &= 2^2 * 324 \\ &= 2^3 * 162 \\ &= 2^4 * 81 \\ &= 2^4 * 3^4 \end{aligned}$$

This means $1296 = 2^4 * 3^4 = 6^4$, and since $1296 = s^4$, s must equal 6. Then, the sum of all the sides must be $6 + 6 + 6 + 6 = \boxed{24}$.

20. If the cost of an apple and 2 oranges is 60 cents, the cost of an orange and 2 peaches is 36, the cost of a peach and 2 apricots is 120, and the cost of an apricot and 2 apples is 96, what is the cost of buying one of each (in cents)?

    Solution: First, set up mathematical equations for the cost of each set of fruits:

$$\begin{gathered} \text{apple} + 2 \text{ oranges} = 60 \\ 2 \text{ peaches} + \text{orange} = 36 \\ \text{peach} + 2 \text{ apricots} = 120 \\ \text{apricot} + 2 \text{ apples} = 96 \end{gathered}$$

    Adding all of these equations together results in the cost to buy three of each fruit: $3 \times (\text{apple} + \text{orange} + \text{peach} + \text{apricot}) = 312$. Then, each side of this equation by 3 to find that the cost to buy one of each fruit is $\boxed{104 \text{ cents}}$.

21. Three animals are arguing about a missing cookie. Frog says that Rabbit ate it, Rabbit says that Frog ate it, and Bird says that Frog is innocent. If exactly 2 animals are lying, who is sure to be innocent?

    Solution: If exactly two of the animals are lying, then one animal must be telling the truth. So, to do this problem, assume one of the animals is telling the truth, and see if there is a possible scenario where the other two animals would be lying.

    If Frog is telling the truth, then Rabbit ate the cookie. Then Rabbit would be lying, because it says Frog ate the cookie, but Bird says Frog is innocent, which is true. This case does not match the criteria because two of the animals are telling the truth. For the criteria to be true, exactly two of the animals must be liars.

    If Bird is telling the truth, then Frog is innocent. However, if Frog is innocent, then either Rabbit or Bird would've eaten the cookie. Since Frog said Rabbit ate the cookie, and he is lying, Bird must have eaten the cookie. This is a possible case since Rabbit would have lied when he said Frog ate the cookie.

If Rabbit is telling the truth, then Frog must have ate the cookie. Then Frog must be lying because it says that Rabbit ate the cookie, and Bird must be lying because it says Frog is innocent. Frog and Bird are both liars, so this case matches the criteria.

This means either Frog or Bird ate the cookie, so $\boxed{\text{Rabbit}}$ is innocent.

22. A man walks on a path around his house. Starting from his house, he first walks 30 feet east, then 40 feet north, then 60 feet west, then finally 80 feet south. How far, in feet, is he from his house?

    Solution: First, looking at just north and south, if he moves 40 feet north and then 80 feet south, he ends up 40 feet south from where he started. Then, looking at only east and west, if he moves 30 feet east and then 60 feet west, the man ends up 30 feet west. Because south and west are perpendicular, they can make a right triangle with legs 30 and 40 feet, as shown in the figure below. Then use the Pythagorean theorem to find the distance:

    $$30^2 + 40^2 = 900 + 1600 = 2500 = distance^2$$

    Then, square rooting both sides results in: $distance = 50$. Therefore, he must be $\boxed{\text{50 feet}}$ from his house.

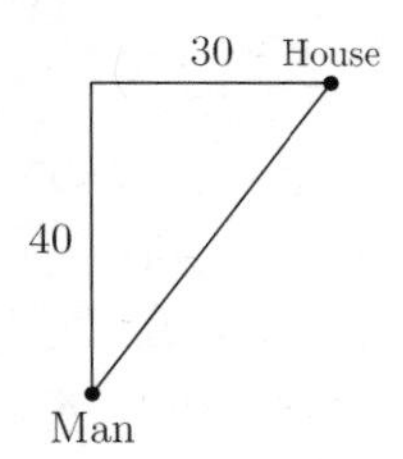

23. John has some cookies. He gives half of them to his siblings, one-third of them to his parents, and one-twelfth to his friend. He eats the only cookie left. How many cookies did he start with?

    Solution: Let's say John has x cookies. He has $x - \frac{x}{2} = \frac{x}{2}$ cookies after he gives them to his siblings, and $\frac{x}{2} - \frac{x}{3} = \frac{x}{6}$ cookies after he shares with his parents. Then, he has $\frac{x}{6} - \frac{x}{12} = \frac{x}{12}$ cookies after he gives some to his friend. Since he had one cookie left after he gave cookies to his family and friends, $\frac{x}{12} = 1$ cookie. Multiplying by 12 on both sides of this equation shows that $x = 12$, so John must have started with $\boxed{\text{12 cookies}}$.

24. If Josh took the square root of a positive number, tripled that number, and then added 8, he got the number 20. But instead of taking the

square root of the number, he should have squared it. If Josh had squared the number, what would his result be?

Solution: Let Josh's positive number be x. If he square rooted this number, multiplied it by three, and added 8, he would have gotten $3\sqrt{x}+8$. To find his original number, set $3\sqrt{x}+8$ to 20, and solve for x, like so:

$$\begin{aligned}3\sqrt{x}+8&=20\\3\sqrt{x}&=12\\\sqrt{x}&=4\\x&=16\end{aligned}$$

However, since he was supposed to square the number instead of square rooting it, his result should have been $3x^2+8$, or:

$$3\times(16^2)+8=3*256+8=768+8=\boxed{776}$$

25. Let $a\wr b=\frac{a|b-a|}{b}$, and $a\ddagger b=a\wr b+b\wr a$. What is $(4\ddagger 6)\wr(6\ddagger 4)$?

    Solution: First, realize $4\ddagger 6=4\wr 6+6\wr 4=6\wr 4+4\wr 6=6\ddagger 4$. This means $4\ddagger 6=6\ddagger 4$, or $(4\ddagger 6)\wr(6\ddagger 4)=(4\ddagger 6)\wr(4\ddagger 6)$. Then, solve for $(4\ddagger 6)\wr(4\ddagger 6)$:

$$\begin{aligned}(4\ddagger 6)\wr(4\ddagger 6)&=\frac{(4\ddagger 6)\times|(4\ddagger 6)-(4\ddagger 6)|}{(4\ddagger 6)}\\&=\frac{(4\ddagger 6)\times 0}{(4\ddagger 6)}=\boxed{0}\end{aligned}$$

26. In the month of May there are 31 days. If May 4th is a Sunday, how many Sundays are there in May?

    Solution: If the 4th is a Sunday, the 11th, 18th, 25th are all Sundays. This means there are a total of $\boxed{\text{4 Sundays}}$.

27. An 8-sided die is rolled (sides numbered 1-8), and then a 6-sided die is rolled. What is the probability that at least 1 prime number is rolled? Express your answer as a common fraction.

    Solution: The easiest way to find the probability of rolling at least one prime number is to find every way this condition will NOT be fulfilled. Then, since the probability of anything happening is 100% (or 1), the probability of rolling at least one prime number will be 1 minus the probability the condition will NOT be fulfilled. In order for this condition NOT to be fulfilled, no prime numbers need can be rolled.

    First, find the odds of NOT rolling a prime number on an 8-sided die. The prime numbers less than or equal to 8 are 2, 3, 5, and 7.

There are four prime numbers out of a total 8 possible outcomes, so the probability of rolling a prime would be 4⁄8=1⁄2. This also means the probability of not rolling a prime is 1⁄2.

Next, find the odds of NOT rolling a prime number on a 6-sided die. The prime numbers less than or equal to 6 are 2,3, and 5. There are three prime numbers out of a total 6 possible outcomes, so the probability of rolling a prime would be 3⁄6=1⁄2.This also means the probability of not rolling a prime is 1⁄2.

Therefore, odds of NOT rolling a prime number is $\frac{1}{2} \times \frac{1}{2} = \frac{1}{4}$. Therefore, the odds of rolling at least one prime number is $1 - \frac{1}{4} = \boxed{\frac{3}{4}}$.

28. Ian is trying to draw a right triangle with a hypotenuse length of 15 and a leg length of 4. How long will the other leg be? Express your answer in simplest radical form.

Solution: Because this is a right triangle, use the Pythagorean theorem: $a^2 + b^2 = c^2$. In this formula, the hypotenuse is c, and the leg can be either a or b. Plugging 15 in for c and 4 for b results in the equation:

$$\begin{aligned} a^2 + 4^2 &= 15^2 \\ a^2 + 16 &= 225 \\ a^2 &= 209 \\ a &= \boxed{\sqrt{209}} \end{aligned}$$

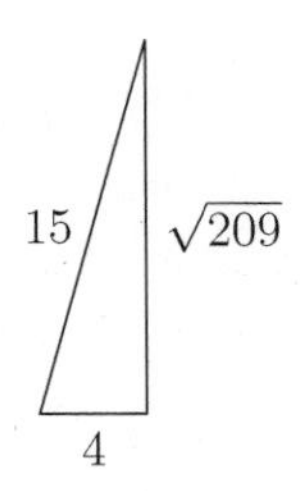

29. The sum of 10 consecutive integers, starting with 11, equals the sum of 5 consecutive integers, starting with what number?

Solution: Let a be the smallest of the 5 consecutive integers. Then, $a + 1$ would be the second integer, $a + 2$ would be the third, and so on. The sum of the 10 consecutive integers from 11 to 20 equals the 5 consecutive integers from a to $a + 4$. Putting this into an equation:

$$11 + 12 + 13 + \ldots + 20 = a + (a + 1) + (a + 2) + (a + 3) + (a + 4)$$

To find the sum of 11 through 20 faster, realize that the sum of 11 and 20, 12 and 19, 13 and 18, 14, and 17, and 15 and 16 all equal 31.

$$\begin{aligned} 11+12+13+14+15+16+17+18+19+20 &= 5a+10 \\ (11+20)+(12+19)+(13+18)+(14+17)+(15+16) &= 5a+10 \\ 31+31+31+31+31 &= 5a+10 \\ 155 &= 5a+10 \\ 145 &= 5a \\ a &= \boxed{29} \end{aligned}$$

30. Annie is playing games of solitaire and listening to music at the same time. Each game of solitaire lasts for 5 minutes. She only wins a game if she hears her special song for some portion of that game. If her special song will play 30 minutes after she begins playing solitaire, Annie will win for the first time when she plays her nth game. What is the value of n?

    Solution: Since it has been 30 minutes, and each game lasts 5 minutes, Annie has played $30 \div 5 = 6$ games. So, the first special song must start with the $\boxed{\text{7th game}}$.

Target Round

1. Sarah is delivering newspapers on a street of houses numbered 1 to 100 with even-numbered houses on the right and odd-numbered houses on the left. If she only delivers to house numbers that are a multiple of 3 and on the right, how many houses does she deliver newspapers to?

   Solution: First, since all numbers on the right are even, then all of these numbers are multiples of 2. Sarah only delivers newspapers to houses that are a multiple of 3 and that are on the right, so that means she delivers newspapers to houses that are a multiple of 3 and of 2.

   In order to be a multiple of 3 and a multiple of 2, the number has to be a multiple of $3 \times 2 = 6$. Therefore, Sarah delivered the newspapers to all houses with numbers that are a multiple of 6. To find the number of houses she delivered to, divide the number of houses (100) by 6 to find the total number of multiples of 6. This results in $100 \div 6 = 16\frac{4}{6}$, but the remainder does not give another multiple of 6, so it can be ignored. Therefore, she delivered newspapers to $\boxed{\text{16 houses}}$.

2. A room is one-half full of people. After 20 people leave, the room is one-third full. How many people would be in the room if it were full?

   Solution: Let c be the number of people in the room if it is full.

   At the start the problem, there are $\frac{1}{2}c$ people because the room is one-half full of people. Then, after 20 people leave, the room is one-third full, which results in the equation $\frac{1}{2}c - 20 = \frac{1}{3}c$.

   Next, isolate the variable by putting both of the terms with variables on one side: $20 = \frac{1}{2}c - \frac{1}{3}c$. To subtract fractions, create a common denominator and simplify the equation: $20 = \frac{3}{6}c - \frac{2}{6}c = \frac{1}{6}c$. Finally, solve for c by multiplying both sides of the equation $\frac{1}{6}c = 20$ by 6 to get $c = 120$. This means the full room capacity is $\boxed{\text{120 people}}$.

3. A cell phone battery is at 40% and will run out in 36 minutes. How long (in minutes) will the battery last if it is at 90%?

   Solution: First, since the phone will last for 36 minutes when the battery is at 40%, then the phone will take $\frac{36\ min}{40\%} = \frac{9}{10}$ of a minute to lose 1% of battery. Then, since the phone is at 90%, the phone will take $90 \times \frac{9}{10} = \boxed{\text{81 minutes}}$ to run out of charge.

4. Paul is not very good at talking to girls. Every time Paul tries to talk to a girl, he gets very nervous and starts counting in multiples of 3.

If Paul counts in multiples of 3 (3, 6, 9, 12, 15,...) until he reaches 1,242, how many numbers has Paul said?

Solution: In this problem, Paul always follows a consistent pattern, and only says multiples of 3. This is because the first number of the sequence is $3 = 1 \times 3$, the second number of the sequence is $6 = 2 \times 3$, the third number is $9 = 3 \times 3$, and so on.

This means that the nth number in the sequence is $n \times 3$. Therefore, to find out how many numbers Paul has said, set $n \times 3$ equal to 1,242. Then, after dividing 1,242 by 3, it is revealed that 1,242 is the 414th number he has said. This means Paul has said a total of $\boxed{\text{414 numbers}}$.

5. In the month of April there are 30 days. How many possible ratios are there of the number of Mondays to the number of Thursdays in April? (April can start on any day of the week.)

   Solution: To solve this problem, count the number of Mondays and the number of Thursdays if we set the first of the month to be Sunday, Monday, and so on until Saturday.

   If the first day is set to be a Sunday or Monday, there we be five Mondays and 4 Thursdays in the month. If the first day is set to be a Tuesday, there will be four Mondays and four Thursdays. If the first day is set to be a Wednesday or Thursday, there will be four Mondays and five Thursdays. Finally, if the first day is a Friday or Saturday, there will be four Mondays and four Thursdays.

   This means there can be five Mondays and four Thursdays, four Mondays and five Thursdays, or four Mondays and four Thursdays. Therefore, there are $\boxed{\text{3 possible ratios}}$.

6. Miki's age is 9 less than 2 times her sister's age. In 4 years, the sum of their ages will be twice their brother's current age, which is 5 more than her sister's current age. What is my age?

   Solution: Let Miki's age be m, her brother's age be b, and her sister's age be s.

   First, set up the equations with their ages. If Miki's age is 9 less than twice her sister's age, then $m = 2s - 9$. Then, in 4 years, the sum of Miki's age and the sister's age will be twice the brother's current age, then $(m + 4) + (s + 4) = 2b$. Finally, since the brothers age is 5 more than his sister's age, then $b = s + 5$.

   Then, since $m = 2s - 9$, and $b = s + 5$, the values of b and m can be substituted into $m + s = 2b$:

$$(m + 4) + (s + 4) = 2b$$
$$(2s - 9 + 4) + (s + 4) = 2(s + 5)$$
$$3s - 1 = 2s + 10$$
$$s = 11$$

Then, the brother's age is $s + 5 = 11 + 5 = 16$ years old, and Miki's age is $2s - 9 = 2(11) - 9 = \boxed{13 \text{ years old}}$.

7. Robert, Amber, Charlie, Jasmine, Annie and Patricia were sitting next to each other (in that order) and one of them is holding a movie. The people at the ends are not holding the movie. The people sitting 2 spaces from either end are not holding the movie. The person holding the movie is no more than 3 spaces away from Patricia. Who is holding the movie?

   Solution: As demonstrated by the visual below, Robert and Patricia who are sitting at the ends, so they are not holding the movie. Next, Charlie and Jasmine are not holding the movie because they are sitting two spots away from the ends. The choices are now Amber and Annie. Amber is four spaces from Patricia, and the person holding the movie cannot be more than three spaces away from Patricia. Therefore, $\boxed{\text{Annie}}$ must be holding the movie.

   Robert Amber Charlie Jasmine Annie Patricia

8. Sally and her boyfriend Jason are going on a romantic picnic. Since Sally is very forgetful, he does not know what kinds of sandwiches Jason enjoys. To accommodate for this, she packs 3 types of breads, 5 types of cheeses (including some stinky ones!), and 6 types of meat. However, Jason refuses to have blue cheese with wheat bread, or white bread with roasted ham. How many choices of sandwiches does Jason have, if he must choose one bread, one cheese, and one meat?

   Solution: First, find the total number of sandwiches. This can be found by multiplying the number of breads, cheeses, and meats, which is $3 \times 5 \times 6 = 90$ choices.

   Next, find the amount of sandwiches that cannot be eaten. If a sandwich has white bread and roasted ham, then there are 5 possible cheeses that can be put on this sandwich. This means there are 5 sandwiches with white bread and roasted ham. Using this same logic, there are 6 sandwiches with wheat bread and blue cheese because there are 6 possible choices for meat.

   This means there are $5+6 = 11$ sandwiches that Jason does not want, which leaves him with $90 - 11 = \boxed{79 \text{ sandwiches}}$ to choose from.

5th-6th Grade 2015 Solutions
Sprint Round

---

1. Susie has a rectangle. If she increases the length by 10% and width by 5%, by what percent is the area increased? Express your answer as a decimal rounded to the nearest tenth.

   Solution: First, realize that the area of a rectangle is equal to the length multiplied by the width ($Area = l \times w$).

   By increasing the length by 10%, it means 10% of the length is added to the original length. Therefore, the new length is:

   $$(100 + 10)\% \times l = 110\% \times l = 1.1 \times l$$

   Using the same method, increasing the width by 5% makes the new width:

   $$(100 + 5)\% \times w = 105\% \times w = 1.05 \times w$$

   With the new width and length, the new area of the rectangle is:

   $$1.1\ l \times 1.05w = 1.155 \times l \times w = 1.155 \times Area = 115.5\% \times Area$$

   If the area of the new rectangle is 115.5% of the original area, the area must have increased by $\boxed{15.5\%}$.

2. Bill has 5 unique Rubik's Cubes to give away to 4 people: Leon, Suhn, Mitchell, and Lucy. How many ways can he distribute them?

   Solution: Since each Rubik's Cube can be given to one of 4 people, there must be 4 ways to hand out each Rubik's cube. Since there are 5 unique Rubik's cubes, and each cube can be handed out 4 ways, there are $4 \times 4 \times 4 \times 4 \times 4$ ways to give out all five Rubik's cubes. $4 \times 4 \times 4 \times 4 \times 4 = 4^5 = 1024$, so there are $\boxed{1024 \text{ ways}}$ to distribute the Rubik's cubes.

3. What is the positive difference between $1 + 2 + 3 + 4 + 5 + \ldots + 100$ and $2 + 3 + 4 + 5 + 6 + \ldots + 101$?

   Solution 1: First, notice is that there are 100 numbers within each set. Also, in the second set, each number is exactly 1 more than the number in the corresponding position in the first set. (For example, the first number of the second set, which is 2, is 1 more than the first number of the first set, which is 1.) Therefore, since each pair of numbers has a difference of 1, and there are 100 pairs, the difference between the two sets is $1 \times 100 = \boxed{100}$.

Solution 2: Let $a = 2 + 3 + 4 + 5 + \ldots + 100$. This means that the first set of numbers is $1 + a$, and the second set of numbers is $a + 101$. Then, subtract $a + 1$ from $a + 101$:

$$a + 101 - (a + 1) = 101 - 1 = \boxed{100}$$

4. A sequence is defined by $a_n = a_{n-1} \times n$. If a_5 is equal to 180, find the value of a_2.

   Solution: In order to solve the problem, work backwards and find a_4, then a_3, and finally a_2.

   To find a_4, plug $n = 5$ into the equation, which results in $a_5 = a_4 \times 5$. After substituting a_5 for 180 and then simplifying, $a_4 = 36$. Next, plug $n = 4$ into the same equation to get a_3.

   Plugging $n = 4$ into the equation results in $a_4 = a_3 \times 4$. After substituting a_4 for 36 and solving, $a_3 = 9$. Then, plug in $n = 3$ to solve for a_2.

   Plugging $n = 3$ into the equation results in $a_3 = a_2 \times 3$, and after substituting a_3 for 9 and solving results in $a_2 = \boxed{3}$.

5. The sides of a right triangle have lengths 8 cm, x cm, and 15 cm, and x is an integer. What is the area of the triangle?

   Solution: In order to solve this problem, use the Pythagorean Theorem. The Pythagorean Theorem states that for each right triangle $a^2 + b^2 = c^2$, where a and b are the lengths of the two legs, and c is the length of the hypotenuse.

   Since x could be less than or greater than 15, it could be the leg or the hypotenuse. Therefore, there are two cases:

   Case 1: x is a leg of the triangle

$$8^2 + x^2 = 15^2$$
$$x^2 = 225 - 64$$
$$x = \sqrt{161}$$

   In this case, x would not be an integer, so x cannot be a leg of the triangle.

   Case 2: x is the hypotenuse

$$8^2 + 15^2 = x^2$$
$$x^2 = 225 + 64$$
$$x = \sqrt{289} = 17$$

   In this case, x would be an integer, so x must be the hypotenuse of the triangle.

For a right triangle, the area is $\frac{1}{2} \times base \times height$, which is equal to $\frac{1}{2}$ times the product of the legs of the triangle. Since x is the hypotenuse, the area of the triangle is $\frac{1}{2} \times 8 \times 15 = 4 \times 15 = \boxed{60}$.

6. In triangle ABC, angle A is equal to 65 and angle B is equal to 75. The bisector of angle C is drawn to meet AB at point D. What is the measure of angle ADC? (An angle bisector is a line that cuts an angle into two angles of equal measure.)

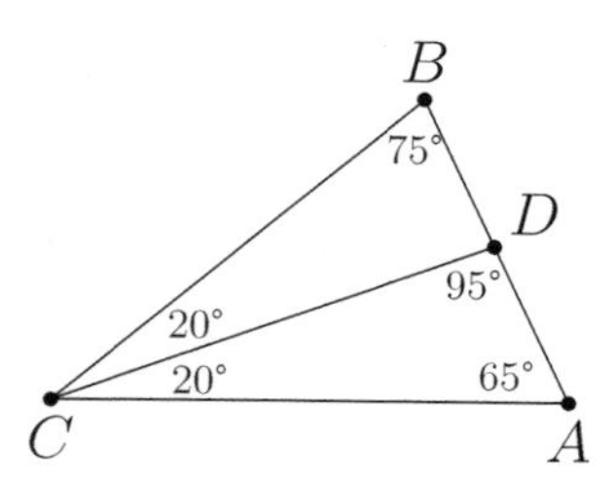

Solution: Before attempting this problem, it helps to know that the sum of the angles in a triangle is equal to 180°. Therefore, find angle C by subtracting 65° and 75° from 180°. This means angle C is equal to:

$$180^\circ - 75^\circ - 65^\circ = 40^\circ$$

If a bisector is drawn through angle C, that means the bisector divides angle C into two equal angles that are each half of angle C, which is $40^\circ \div 2 = 20^\circ$. So, if angle A is equal to 65°, and angle ACD is equal to 20°, then angle ADC must equal $180^\circ - 65^\circ - 20^\circ = \boxed{95^\circ}$.

7. Five friends, Natalie, Lucy, Amy, Thomas, and Ian, go to Disneyland together. They each buy their own ticket for \$156 each. They equally split up the lunch bill of \$100 and they each buy one pair of Mickey Mouse ears for \$22 a pair. If this is all they spent, how many dollars did each of them spend at Disneyland?

Solution: Each of their tickets cost \$156. If they split up the lunch bill among the 5 of them, then they each spent $\$100 \div 5 = \$20$ on lunch. And finally, they each bought a pair of Mickey Mouse ears for \$22. So, each person spent a total of $\$156 + \$20 + \$22 = \boxed{\$198}$.

8. William, Zack, Bryan, Jason, and Joey are running in a race. These statements are true about the results:

William finished before Jason but after Joey.

Jason finished in third place.

Bryan finished after Joey but he did not finish last.

Zack finished after both Bryan and Jason.

Who was the first place runner?

Solution: With these types of questions, it helps to draw a diagram similar to the one below. Imagine that they are in a line, and the person on the left side of the line is the one who finished first, and the person on the right side of the line is the one who finished last. The most specific piece of information that is given is that Jason finished in third place. This means he can be placed in the middle:

______ ______ Jason ______ ______

Next, William finished before Jason, but after Joey. Since there are only two spots in front of Jason, Joey must have been the first to finish, followed by William:

Joey William Jason ______ ______

Next, Zack finished after Bryan and Jason. Because there are only two spots left after Jason, this must mean Zack finished last, and Bryan finished second to last:

Joey William Jason Bryan Zack

Therefore, $\boxed{\text{Joey}}$ was the first place runner. (Note: The last paragraph of information is not necessary to find the answer to the problem, but it helps complete the entire diagram.)

9. If the lengths of two sides of a isosceles triangle are 5 and 8 respectively, what is the sum of all the possible perimeters for the triangle?

   Solution: An isosceles triangle is a triangle with two sides that are equal. So, if two of the sides are 5 and 8, then there are two possible triangles: one with sides 5, 5, and 8, and the other with sides 5, 8, and 8.

   Before assuming these are both answers, make sure triangles with those sides are possible. Within a triangle, the sum of any two sides must always be greater than the third remaining side. Luckily, $5+5 > 8$ and $5+8 > 8$, so both triangles are possible solutions.

   The perimeter for the first possible triangle is $5+5+8 = 18$ and the perimeter for the second possible triangle is $5+8+8 = 21$. Therefore, the sum of these all possible perimeters is $18 + 21 = \boxed{39}$.

10. Bob sells gum in packs of 3 and 5. What is the largest amount of gum that a person is not able to buy using a combination of Bob's packs of gum? For example, a person can get 14 pieces of gum by ordering 3 packs of 3 and 1 pack of 5.

Solution: Let n represent any non-negative integer. First off, since the gum is sold in packs of 3, a person can buy $3n$ (meaning any multiple of 3) sticks of gum. Also, they can buy $5+3n$ sticks of gum by buying one pack of 5 and n packs of 3. In addition, they can buy $10+3n$ sticks of gum by buying two packs of 5 and n packs of 3.

Notice that for any number greater than 10, a person can buy that many pieces of gum by using one of the three methods above. This means that the greatest amount of pieces someone cannot buy must be less than 10. Then, plug in numbers starting from 9. It is possible to get 9 pieces using 3 packs of 3, and to get 8 pieces using a pack of 3 and a pack of 5. However, it is not possible to get 7 pieces of gum. Therefore, the answer is $\boxed{\text{7 pieces of gum}}$.

11. Susan is a criminal case solver. Today, her case is to find out who ate which type of candy, Skittles, M&M's, Jolly Ranchers, and Twix. She knows Michael hates chocolate. Mindy says she would never eat Twix in a million years. John said that he saw a boy eat Skittles. Lastly, Joey is under a lot of pressure, and confesses that he ate the Jolly Ranchers. Given that nobody ate the same candy, what did Mindy eat? (Assume M&M's and Twix are the only two chocolate candies.)

    Solution: To start off, list out all of the suspects who ate candy: Joey, Michael, Mindy, and John. Joey must have eaten Jolly Ranchers from his confession. Then, John said he saw a boy eat Skittles, and since Michael hates chocolate and did not eat the Jolly Ranchers, Michael must have eaten the Skittles. Since John and Mindy are the only people who haven't been assigned a candy, one of them must have eaten Twix, and the other must have eaten M&M's. Mindy said she would never eat Twix in a million years, so she must have eaten the $\boxed{\text{M\&M's}}$.

12. Leon is playing Tic-Tac-Toe with Marcia Dai. Each game takes 16 minutes and 40 seconds. Assuming they don't take breaks how many games will they play in ten days?

    Solution: First, the number of minutes to seconds. Since there are 60 seconds in a minute, 16 minutes and 40 seconds is equal to $16 \times 60 + 40 = 1000$ seconds. Next, calculate how many seconds are in 10 days. (Note that there are 24 hours in a day, 60 minutes in an hour, and 60 seconds in a minute.)

    $10\ days \times 24\ \frac{hours}{day} \times 60\ \frac{minutes}{hour} \times 60\ \frac{seconds}{minute} = 864,000$ seconds. Since each game is 1,000 seconds, and they have a total of 864,000 seconds to play, then they can play $864,000/1,000 = \boxed{\text{864 games}}$.

13. Using a standard deck of cards, a class is doing a fun game. When a certain suit is a pulled from the deck, the class has to do a certain

exercise. Hearts are jumping jacks, diamonds are sit-ups, clubs are push-ups, and spades are burpies. If all face cards (including aces) represent 10, and the number on the card represent how many of each exercise you need to do, how many sit-ups and push-ups would a student have completed after going through the whole deck?

Solution: Realize that there are 52 cards in a standard deck, and there are 13 of each suit: hearts, diamonds, clubs, and spades. Note that the face cards (King, Queen, Jack and Ace) each mean the class must do 10 of a particular exercise. So, the number of sit-ups and push-ups a student needs to do is two times the sum of one suit of cards. The sum of one suit equals $2+3+4+5+6+7+8+9+10+(4\times 10)$, which equals 94. So, the total number of sit-ups and push-ups equals $2\times 94 = \boxed{188}$.

14. If $x = 3y$ and $z = \frac{x}{2}$, find the value of $\frac{x+y+z}{x+y-z}$

Solution: First, notice that since $z = \frac{x}{2}$, every z in the expression $\frac{x+y+z}{x+y-z}$ can be replace by $\frac{x}{2}$. Doing this results in this expression:

$$\frac{x+y+\frac{x}{2}}{x+y-\frac{x}{2}}$$

Similarly, since $x = 3y$, every x in the expression can be replaced by $3y$. Doing this results in this expression:

$$\frac{3y+y+\frac{3y}{2}}{x+y-\frac{3y}{2}} = \frac{\frac{11y}{2}}{\frac{5y}{2}} = \frac{11\times\frac{y}{2}}{5\times\frac{y}{2}} = \boxed{\frac{11}{5}}$$

15. With the letters I N D I A N A, how many ways are there to order the letters?

Solution: First, pretend that the word INDIANA consisted of 7 different letters (so no two letters are the same). If this were the case, then it is easy to calculate the number of ways to order the letters. This would mean that 7 possible letters could be the first letter of the new arrangement. Once the first letter is picked, there are 6 possible letter that could be the second letter, which would leave 5 possible letters for the third letter, 4 possible letters for the fourth letter, and so on. Then, multiply the possible number of letters in each position to get $7\times 6\times 5\times 4\times 3\times 2\times 1 = 5040$ possible combinations.

$$\underline{\ 7\ } \times \underline{\ 6\ } \times \underline{\ 5\ } \times \underline{\ 4\ } \times \underline{\ 3\ } \times \underline{\ 2\ } \times \underline{\ 1\ } = 5040$$

However, the word INDIANA does not have 7 different letters. Instead, it has 2 I's, 2 A's, and 2 N's. In any particular arrangement, switching the position of a pair of two of the same letters results in the same arrangement. This means that for every pair of letters, the number of combinations must be divided by 2. This makes the true number of arrangements $5040 \div 2 \div 2 \div 2 = \boxed{630}$ arrangements.

16. If 2 cans of paint are enough to paint a 1 yard x 1 yard square, how many cans are needed to paint the outside walls of a shed that is 10 yards long, 8 yards tall, and 4 yards wide? (The roof and ground do not need painting)

    Solution: The first step in solving would be to find the surface area of the walls of the shed, not including the roof or the ground. Two of the walls will have an area of $10 \times 8 = 80\ yards^2$, and the other two walls will have an area of $4 \times 8 = 32\ yards^2$. Then, add the area of all four walls together to get the total area that needs to be painted:

    $$80 + 80 + 32 + 32 = 224\ yards^2$$

    Then, since each square yard takes two cans of paint to cover, and the area of the walls is $224\ yards^2$, it will take $224\ yards^2\ \times 2\ \frac{cans}{yard^2} =$ $\boxed{448 \text{ cans}}$ of paint to cover the walls.

17. The Apple County Math Circle is choosing a committee of a president, vice president, and a secretary from a group of 30 members. How many committees can be made?

    Solution: For this problem, observe that there are 3 positions to be filled by 30 possible candidates. Therefore, starting from president, there are 30 eligible people that can become president. Once a president is chosen, there will only be 29 people who can be vice president. Lastly, once a vice president is chosen, there will only be 28 people available for secretary.

    To find the number of committees, multiply the possible number of presidents by the possible number of vice presidents by the possible number of secretaries. Calculating this results in $30 \times 29 \times 28 = 570 \times 28 = \boxed{23{,}460 \text{ possible combinations}}$.

18. A store offers a free bucket of paint for every 3 buckets bought. Robert buys 10 buckets of paint, and Bob buys 6 buckets of paint. If each bucket of paint costs \$10 each, how much money is saved if they bought the paint together rather than separately?

    Solution: First, find out how much they each paid separately. Robert bough 10 buckets, and since he got a free bucket after purchasing three, he payed for 8 buckets, and got two free, making his individual total $8 \times 10 =$\$80. Bob bought six buckets, and because of the deal, he only paid for five buckets and got one for free, paying $5 \times 10 =$\$50. Adding those up, if paying individually, they paid a total of \$130.

If they bought the buckets together, they would need to buy 16 buckets of paint, which would allow them to pay for twelve, and get four for free, resulting in $12 \times 10 =$ \$120.

Therefore, the total amount saved is $\$130 - \$120 = \boxed{\$10}$.

19. In a party with 10 people, each person shakes hands once with everybody else. How many total handshakes occur?

    Solution: To find the amount of handshakes possible, multiply the amount of people able to handshake (Person A) by the amount of people left to reciprocate the handshake (Person B). There are 10 people, and 9 people to shake with, so the number of handshakes is $10 \times 9 = 90$.

    However, since Person A shaking hands with Person B is the same as Person B shaking hands with Person A, our calculations have counted each handshake twice. Therefore, divide the number of handshakes by two to find the total number of handshakes that occurred: $90 \div 2 = \boxed{45 \text{ handshakes}}$.

20. The perimeter of an equilateral triangle is equal to that of a rectangle. If the side length of the triangle is 12, what is the maximum area the rectangle can have?

    Solution: In order to find the side lengths of the rectangle, first find the perimeter of the triangle. Since each side is 12, and it is an equilateral triangle, all three sides are equal. This makes the perimeter of the triangle and the rectangle $12 \times 3 = 36$.

    Next, in order to find the largest area of the rectangle, notice that the length and width of the rectangle must be as close as possible. Then, since 36 is divisible by 4, the rectangle with the closest length and width is a square with side length $36 \div 4 = 9$. The area of this square is the side length of the square multiplied by itself, which is $9 \times 9 = \boxed{81}$.

21. If $x + \frac{1}{x} = 10$, calculate $x^2 + \frac{1}{x^2}$.

    Solution: For this problem, the left side of the first equation needs to look like what the problem asks for. To do this, the first step is to square both sides of the first equation:

$$(x + \frac{1}{x})^2 = 10^2$$

Simplifying this expression results in:

$$x^2 + 2(x * \frac{1}{x}) + \frac{1}{x^2} = 100$$
$$x^2 + 2 + \frac{1}{x^2} = 100$$
$$x^2 + \frac{1}{x^2} = \boxed{98}$$

22. Find $x+y$ if both x and y are whole numbers and $(2x+y)(x-2y) = 7$ is satisfied.

    Solution: Since both x and y are both whole numbers, both $(2x + y)$ and $(x - 2y)$ must also be whole numbers. Since 7 is a prime number, the only two positive numbers that multiply to get 7 are 1 and 7. Therefore, the bigger number, $(2x+y)$, is 7, and the smaller, $(x-2y)$, is 1.

    This results in a system of equations: $2x + y = 7$ and $x - 2y = 1$. Adding $2y$ to both sides of the second equation results in $x = 2y+1$. Then, substitute $2y + 1$ for x in the first equation:

$$2x + y = 7$$
$$2(2y + 1) + y = 7$$
$$4y + 2 + y = 7$$
$$5y = 5$$
$$y = 1$$

    Then, solve for x using the modified first equation, and find $x + y$.

$$x = 2(1) + 1 = 3$$
$$x + y = 3 + 1 = \boxed{4}$$

23. They are having student elections at Sophie's school. There are 4 candidates for presidents, 3 candidates for vice president, and 5 candidates for activities commissioner. There were 3 candidates for historian, but one of them decided to opt out of the election, and two more joined in. How many different student bodies can there be?

    Solution: This question is asking us how many different ways the possible candidates for each position can be arranged such that a different student body is created each time. To do this, multiply the number of people that are available for each position:

$$4\ (presidents) \times 3\ (VPs) \times 5\ (ACs) \times 4\ (historians) = \boxed{240}.$$

24. What is the obtuse angle created by the hour hand and minute on an analog clock when it is 3:40?

Solution: First, realize that for each hour, the hour hand of a clock moves $\frac{1}{12}$ of its way around the circle, which is $360 \times \frac{1}{12} = 30$ degrees. When it is 3:00, the hour hand has moved $3/12$ of the way around the circle, or $30 \times 3 = 90$ degrees.

Since it is 40 minutes past that, the hour hand must have moved more than 90 degrees. Realize that 40 minutes is equal to $\frac{40}{60} = \frac{2}{3}$ of an hour. So, the hour hand must have moved another $30 \times \frac{2}{3} = 20$ degrees. This means the hour hand is at $90 + 20 = 110$ degrees.

Then, realize that for each minute, the minute hand of a clock moves $\frac{1}{60}$ of its way around the circle, which is $360 \times \frac{1}{60} = 6$ degrees. Since the minute hand is on the 40 minute mark, the minute hand is $6 \times 40 = 240$ degrees past the 0 minute mark.

The angle between the minute hand and the hour hand is then the difference between the two angles, which is $240 - 110 = \boxed{130°}$.

25. If Matthew Tang destroys a card deck in 5 days while Arthur Liu destroys 3 decks in a week. How many days will it take for them to destroy 22 decks if they work together?

    Solution: Since Arthur destroys 3 decks a week, he must destroy $\frac{3}{7}$ decks a day. Matthew can destroy a deck in 5 days, meaning he destroys $1/5$ a deck every day. If they work together, they can destroy $\frac{3}{7} + \frac{1}{5} = \frac{15}{35} + \frac{7}{35} = \frac{22}{35}$ decks a day. In order to destroy 22 decks, it would take them $22\ decks \div \frac{22}{35} \frac{decks}{day} = \boxed{35 \text{ days}}$.

26. All terms in the following geometric sequence are positive integers. The first three terms are n, n + 3, and 2n + 6. What is the fourth term of this sequence?

    Solution: In order to find the next number in a geometric sequence, the term before it is always multiplied by a certain number r. By looking at the second and third terms, notice that $r \times (n+3) = 2n+6$. Since $2n + 6 = 2 * (n + 3)$, r must equal 2.

    Then, by looking at the first and second terms, notice that $r \times n = 2 \times n = n + 3$. Then, subtracting n from both sides of the equation results in $n = 3$. This means the first term is 3, the second term is $3 \times 2 = 6$, the third term is $6 \times 2 = 12$, and the fourth term is $12 \times 2 = \boxed{24}$.

27. We are playing chess. I am really bad, so I only have my bishop left, and it can only move on the black squares. The chess board that we are playing with is 6×6. What fraction of the board can the bishop not move on?

    Solution: Every other square is a different color, going back and forth between white and black. Since there are only two colors, that means half of the squares will be black, and half will be white, meaning the bishop cannot move on $\boxed{\frac{1}{2}}$ of the board.

28. What is the maximum number of regions that 7 chords can make in a circle?

    Solution: The easiest way to visualize this is to draw this on a circle. When 1 chord is drawn in a circle, it divides the circle into 2 regions. If a second chord is drawn, it can cut through both regions, creating a total of 4 regions. When the third chord is drawn, it can cut through at most 3 of the four regions, creating a total of $4 + 3 = 7$ regions. Then, when the fourth chord is drawn, it can cut through at most four regions, creating a total of $7 + 4 = 11$ regions. As demonstrated by the second, third, and fourth chord, the number of regions that can be cut is the number of the chord being drawn. This means the fifth chord can cut at most five regions, the sixth chord can cut at most six regions, and the seventh chord can cut at most seven regions. This makes the total number of regions $2+2+3+4+5+6+7 = \boxed{29}$.

29. Let $ABCDEF$ be a 6-digit number, and let A, B, C, D, E, F be digits of different values. If $ABCDEF * 4 = EFABCD$, what is $ABCDEF$? (Assume A and E cannot equal 0.)

    Solution: First, notice that using place values, any number can be expressed in terms of powers of 10. For example, the number 3,564 can be rewritten as $1000 * 3 + 100 * 5 + 10 * 6 + 4 * 1$.

    Using this method, rewrite the numbers $ABCDEF$ and $EFABCD$. This makes

$$ABCDEF = 100,000*A+10,000*B+1,000*C+100*D+10*E+F$$

and

$$EFABCD = 100,000*E+10,000*F+1,000*A+100*B+10*C+D$$

Then, $ABCDEF * 4 = EFABCD$ can be expressed as:

$$4 * (100,000 * A + 10,000 * B + 1,000 * C + 100 * D + 10 * E + F)$$
$$= 100,000 * E + 10,000 * F + 1,000 * A + 100 * B + 10 * C + D$$

Then, simplify by distributing the 4 to all variables:

$$400,000 * A + 40,000 * B + 4,000 * C + 400 * D + 40 * E + 4 * F$$
$$= 100,000 * E + 10,000 * F + 1,000 * A + 100 * B + 10 * C + D$$

Next, move each variable to one side of the equation, and factor out any common factors:

$$\begin{aligned} 399,000A + 39,900B + 3,990C + 399D &= 99,960E + 9,996F \\ 399(1000 * A + 100 * B + 10 * C + D) &= 9,996(10 * E + F) \\ 19 * 21(1000 * A + 100 * B + 10 * C + D) &= 476 * 21(10 * E + F) \\ 19(1000 * A + 100 * B + 10 * C + D) &= 476(10 * E + F) \end{aligned}$$

Then, since 19 and 476 do not share any common factors, then $(1000 * A + 100 * B + 10 * C + D)$ must be a multiple of 476, and $(10*E+F)$ must be a multiple of 19. However, since A is the start of a number, it cannot be 0. Therefore, set $(1000*A+100*B+10*C+D)$ to the smallest multiple of 476 that is greater than 1,000: $476*3 = 1,428$. In this case, $A = 1$, $B = 4$, $C = 2$, and $D = 8$.

Then, in order to keep both sides of the equation equal, $(10 * E + F)$ must be $3 * 19 = 57$. This would make $E = 5$, and $F = 7$. Luckily, A, B, C, D, E, F are different digits, so this solution is correct. Therefore $ABCDEF = \boxed{142{,}857}$

30. The length of one leg of a right triangle is 2 inches less than the length of the other leg. If the length of the hypotenuse is 10 inches, what is the length of the longer leg?

    Solution: In this problem, it is useful to know about the 3-4-5 right triangle, which states that any triangle that has a side ratio of 3-4-5 is a right triangle. One possible triangle that follows this principle is the 6-8-10 triangle. This fits the description in the problem since 10 is the hypotenuse and the difference between the legs is 2. Therefore, the length of the longer leg is $\boxed{\text{8 inches}}$.

Target Round

---

1. What is the sum of the first 100 even natural numbers?

   Solution: First of all, realize that even natural numbers start from 2, and continue on (2,4,6,8, 10...). In other words, the problem is asking us what 2+4+6+8+...+200 is equal to.

   In order to solve this problem, visualize these 100 numbers as 50 pairs of numbers. Take the first and last number (2 and 200) and then the second and second to last number (4 and 198) and so on. Notice that each of these pairs add up to 202.

   Since there are 50 of these pairs, the solution will be 202×50, which is $\boxed{10{,}100}$.

2. The cheese pizza at Cheesecake Factory was \$12. For every topping I add, it costs an extra \$1.20. I added pineapple, onions, and pine nuts. After eating my pizza, I gave the waiter 5% tip. If I used a gift card for \$10, how much was my final bill (in dollars)? Express your answer as a decimal rounded to the nearest hundredth.

   Solution: For this problem, the first step would be to calculate what the total amount they paid for the pizza was. First, calculate the prize of the price of the pizza with toppings without tip. Since each topping costs \$1.20, and they got three toppings, the price becomes:

   $$\$12 + (3 \times \$1.20) = \$15.60$$

   \$15.60 is the amount they paid before the tip. In order to calculate the tip, multiply \$15.60 by 5%, or $^5/_{100}$, and add it to the total cost.

   $$(0.05 \times \$15.60) + \$15.60 = \$0.78 + \$15.60 = \$16.38$$

   Then, since they used a \$10 gift card, subtract \$10 from the total to find the value of the final bill: $\$16.38 - \$10.00 = \boxed{\$6.38}$.

3. A strange clock chimes 1 time at 1 o' clock, 2 times at 2 o' clock, 3 times at 3 o' clock, 4 times at 4 o' clock, 5 times at 5 o' clock, and so on. How many times will the clock chime from 1-12 o' clock?

   Solution: If the clock chimes 1 time at 1 o' clock, 2 times at 2 o' clock, 3 times at 3 o' clock and so on, then the number of times the clock will chime from 1-12 o'clock equals:

   $$\begin{aligned} 1+2+3+4+5+6+7+8+9+10+11+12 \\ = (1+2+3+4)+(5+6+7)+(8+12)+(9+11)+10 \\ = 10+18+20+20+10 \\ = \boxed{78 \text{ times}} \end{aligned}$$

4. Using the numbers 1, 2, 3, 5, 8, how many three-digit, even numbers can be formed? (Digits cannot be repeated.)

   Solution: In order to find the number of three digit even numbers that can be formed, find the amount of numbers that can go in each one of the three digits.

   First, deal with the first restriction, and make sure the number is even. In order to fulfill this, the last digit, or the ones digit, must be either 2 or 8. This gives us two different options for the ones digit.

   Then, there are no restrictions, so any number can go in the hundreds and tens digit. However, since numbers cannot be repeated, and one of the five digits was used in the ones digit, that leaves 4 possible options for the tens digit. Using this same logic, this leaves 3 possible options for the hundreds digit.

   To find the solution to the problem, simply multiply the amount of numbers that can be placed in each digit. This results in $2 \times 4 \times 3 = \boxed{24}$.

5. Given that x, y, and z are positive numbers such that $x^2/2 = 4y$ and $8y = z^8$, what is x in terms of z?

   Solution: The easiest way to solve this problem is to multiply the first equation by 2. Doing so will result in $x^2 = 8y$. Then, since $8y = z^8$, replace $8y$ with z^8 in the second equation to get $x^2 = z^8$. Then, square root both sides, which results in $x = \boxed{z^4}$. Note that $x \neq -z^4$ because x and z both must be positive.

6. If the number of seconds in 6 weeks can be expressed as $x!$, find x.

   Solution: To begin this problem, first find the amount of seconds there are in 6 weeks. There are 7 days in a week, 24 hours in a day, 60 minutes in an hour, and 60 seconds in a minute. This means there are this many seconds in a day:

$$6\ weeks \times 7\ \frac{days}{week} \times 24\ \frac{hours}{day} \times 60\ \frac{minutes}{hour} \times 60\ \frac{seconds}{hour}$$

   The problem is asking us to represent the number in terms of $n!$ [Note: $n!$, or n factorial, equal $1 * 2 * 3 * ... * (n-1) * (n)$], so instead of multiplying the number out, try to put the number above in the form: $[1 * 2 * 3 * ... * (n-1) * (n)]$.

$$\begin{aligned} &6 \times 7 \times 24 \times 60 \times 60 \\ &= 6 \times 7 \times (8 \times 3) \times (3 \times 4 \times 5) \times (3 \times 10 \times 2) \\ &= 6 \times 7 \times 8 \times 3 \times 3 \times 4 \times 5 \times 3 \times 10 \times 2 \\ &= 6 \times 7 \times 8 \times 9 \times 4 \times 5 \times 3 \times 10 \times 2 \\ &= 10 \times 9 \times 8 \times 7 \times 6 \times 5 \times 4 \times 3 \times 2 \times 1 = 10! \end{aligned}$$

Therefore, $x = \boxed{10}$.

7. Find the area of a triangle with vertices at (2, 2), (3, 5), and (-1, 3).

   Solution: Let A = (2,2), B = (3,5), and C = (-1,3). Then, draw the diagram on a graph like the one below:

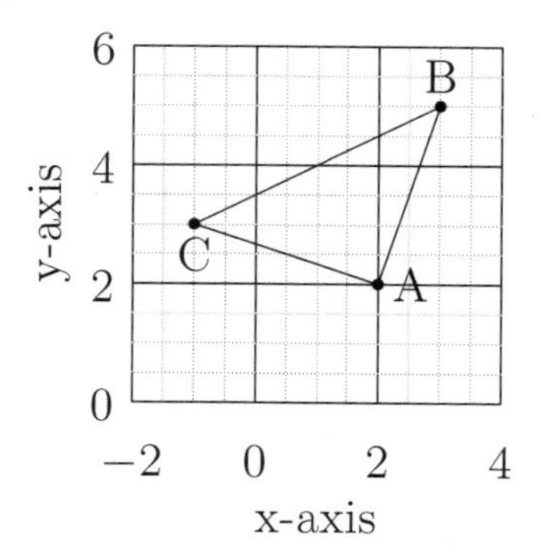

   From this diagram, $\overline{AB}$ and $\overline{AC}$ look perpendicular to one another. To prove this, find the slope of each line, and if they are negative reciprocals of each other, then they are perpendicular:

   slope of $\overline{AB}$: $\frac{5-2}{3-2} = 3$

   slope of $\overline{AC}$: $\frac{3-2}{-1-2} = -\frac{1}{3}$. Therefore, $\overline{AB} \perp \overline{AC}$.

   This means that $\overline{AB}$ and $\overline{AC}$ are the legs of a right triangle. This means that the area can be found by multiplying the base (AC) by the height (AB) by ½. To find the length of $\overline{AB}$ and $\overline{AC}$, use the distance formula (which is a modification of the Pythagorean Theorem) to find the distance between two points on a coordinate plane. The formula is: $\sqrt{(x_2 - x_1)^2 + (y_2 - y_1)^2}$. This means $AB = \sqrt{(3-2)^2 + (5-2)^2} = \sqrt{10}$, and $AC = \sqrt{(-1-2)^2 + (3-2)^2} = \sqrt{10}$. Then, the area must be $\sqrt{10} \times \sqrt{10} \times 1/2 = 10 \div 2 = \boxed{5}$.

8. The lengths of three sides of a triangle are 1027, 2014, and x. If the possible values of x are expressed as $m < x < n$, what does $m + n$ equal?

   Solution: In order to answer this question, one must know the Triangle Inequality Theorem. It states that the length of a side of a triangle will always be less than the sum of the lengths of the other two sides.

   Therefore, realize that $x < 1027 + 2014 = 3041$

   Using the same theorem, realize that 2014 has to be less than the sum of x and 1027. From this the equation becomes $2014 < x + 1027$. This can be simplified to: $2014 - 1027 = 987 < x$.

Combining the two inequalities gives this: $987 < x < 3041$. Therefore $m = 987$, $n = 3041$, and $m + n = 987 + 3041 = \boxed{4028}$.

5th-6th Grade 2016 Solutions

Sprint Round

---

1. You start with 20 apples. You eat half of your apples. Your friend gives you 6 more apples. How many apples do you have?

   Solution: At the beginning, you start with 20 apples. Then, if you eat half of your apples, you have eaten $20 \div 2 = 10$ apples. If you have eaten 10 out of 20 apples, you have $20 - 10 = 10$ apples remaining. Then, once your friend gives you 6 more apples, the final amount of apples is $10 + 6 = \boxed{16 \text{ apples}}$.

2. A parallelogram has an area of 15 cm^2 and height of 7.5 cm. What is the length of its base?

   Solution: First, note that the area of a parallelogram equals its base times its height ($Area = base \times height$).

   Plug in the given area and height of the parallelogram to get $15 = base \times 7.5$. Divide both sides of the equation by 7.5 to get that the base must be $\boxed{2 \text{ cm}}$ long.

3. $3x - 4y = 18$ and $4x - 3y = 10$. Compute $x - y$.

   Solution 1: If these two equations are added together, then:

$$\begin{array}{r} 3x - 4y = 18 \\ +4x - 3y = 10 \\ \hline 7x - 7y = 28 \end{array}$$

   After that, divide both sides of the equation by 7 to get $x - y = \boxed{4}$.

   Solution 2: Another possible way to solve this problem is by calculating x and y using substitution. First, using the equation $3x - 4y = 18$, discover what x is in terms of y:

$$3x - 4y = 18$$
$$3x = 4y + 18$$
$$x = \frac{4}{3}y + 6$$

Then, $\frac{4}{3}y + 6$ can be substituted for x in the equation $4x - 3y = 10$:

$$\begin{aligned} 4x - 3y &= 10 \\ 4(\frac{4}{3}y + 6) - 3y &= 10 \\ \frac{16}{3}y + 24 - 3y &= 10 \\ \frac{7}{3}y &= -14 \\ \frac{1}{3}y &= -2 \\ y &= -6 \end{aligned}$$

Then, since $x = \frac{4}{3}y + 6$, x must be

$$\frac{4 \times (-6)}{3} + 6 = -2$$

This means $x - y = (-2) - (-6) = \boxed{4}$

4. A square and an equilateral triangle have side length 5. Michelle glues one edge of the equilateral triangle to one edge of the square, keeping the square and the triangle touching at only one edge. What is the perimeter of this new figure?

   Solution: First, draw a diagram similar to the one below:

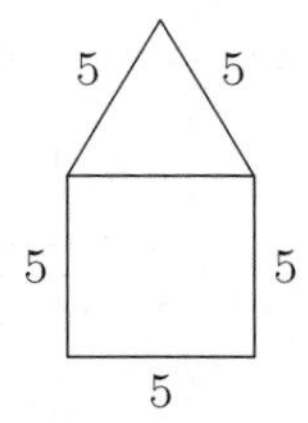

   Then, add up all of the outer sides of the figure to get the perimeter of the house: 5+5+5+5+5= $\boxed{25}$.

5. How many lines of symmetry does the following figure have?

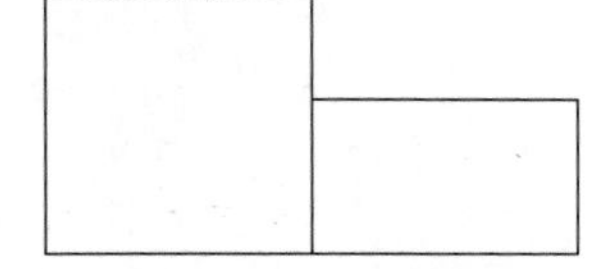

   Solution: If you draw any line through the figure, the two sides will always look different. Therefore, the answer is $\boxed{\text{0 lines of symmetry}}$. (Some example lines of symmetry that don't work are shown on the figure below.)

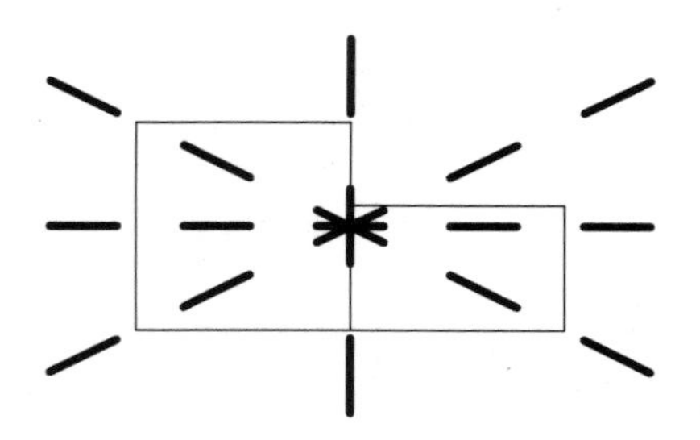

6. The angles of one triangle are $2x°$, $3y°$, and $80°$. The angles of another triangle are $120°$, $x°$, and $y°$. What is $x + y$?

   Solution: First, it is important to know that the sum of the angles in a triangle equals $180°$. This means that $2x + 3y + 80 = 180$, or $2x + 3y = 100$, and that $120 + x + 2y = 180$, or $x + 2y = 60$.

   Then, subtract $2y$ from each side of the second equation to get $x = 60 - 2y$. From this point, this problem can be solved using substitution:

$$\begin{aligned} 2x + 3y &= 100 \\ 2 \times (60 - 2y) + 3y &= 100 \\ 120 - 4y + 3y &= 100 \\ 120 - y &= 100 \\ 120 - 100 = y &= 20 \end{aligned}$$

   Then, since $x = 60 - 2y$,

$$x = 60 - 2 \times 20 = 20$$

   Therefore $x + y$ must equal $20 + 20 = \boxed{40°}$.

7. Clarisse flips a coin twice. What is the probability that the two flips yield different results?

   Solution: There are four possible results when flipping a coin twice: two heads, a head and a tail, a tail and a head, or two tails. 2 of the 4 possibilities give different results, so the answer is $\frac{2}{4} = \boxed{\frac{1}{2}}$.

8. Suppose that $a \diamond b = \frac{ab}{a+b}$. Compute $2 \diamond (3 \diamond 6)$.

   Solution: First, realize that according to order of operations, operations in parentheses must be solved first. This means that in the equation $2 \diamond (3 \diamond 6)$, $(3 \diamond 6)$ must be solved first.

$$\begin{aligned} 2 \diamond (3 \diamond 6) &= 2 \diamond (\frac{3 \times 6}{3 + 6}) = 2 \diamond 2 \\ 2 \diamond 2 &= \frac{2 \times 2}{2 + 2} = \boxed{1} \end{aligned}$$

9. A school has some books. All the books can be split evenly among 6 classrooms and among 10 classrooms. What is the smallest number of books that the school can have?

   Solution 1: If all the books can be split evenly among 6 classrooms, the total number of books must be a multiple of 6. Likewise, the total number of books is also a multiple of 10. To find the least common multiple of 6 and 10, list the multiples of 6 and 10 until both of them share a number.

   Multiples of 6 are: 6, 12, 18, 24, **30**, 36, ...
   Multiples of 10 are: 10, 20, **30**, 40, ...

   The first multiple shared by 6 and 10 is 30. Therefore, the answer is $\boxed{\text{30 books}}$.

   Solution 2: If all the books can be split evenly among 6 classrooms, the total number of books must be a multiple of 6. Likewise, the total number of books is also a multiple of 10. Then, the answer to this problem is the least common multiple of 6 and 10. Since $6 = 2 \times 3$ and $10 = 2 \times 5$, the least common multiple of 6 and 10 is $2 \times 3 \times 5 = \boxed{\text{30 books}}$.

10. Allison has to solve 15 math problems for homework. 5 are easy problems, 5 are medium problems, and 5 are hard problems. If it takes her 2 minutes to solve an easy problem, 4 minutes to solve a medium problem, and 7 minutes to solve a hard problem, how many minutes does it take her to finish her homework?

    Solution: If it takes Allison 2 minutes to solve one easy problem, then it will take her $2 \times 5 = 10$ minutes to solve all 5 easy problems. Likewise, if it takes 4 minutes to solve a medium problem, it will take $4 \times 5 = 20$ minutes to solve all 5 medium problems, and if it takes her 7 minutes to solve one hard problem, it will take her $7 \times 5 = 35$ minutes to solve all 5 hard problems.

    Therefore, it will take $10 + 20 + 35 = \boxed{\text{65 minutes}}$ to solve all 15 problems.

11. Two chipmunks can build a treehouse in five hours. How much time will it take for ten chipmunks to build twenty treehouses?

    Solution 1: If 2 chipmunks can build 1 treehouse in 5 hours, then 2×5 chipmunks (or ten chipmunks) can build 1×5 treehouses in 5 hours. To build 20 treehouses (or 5 treehouses four times), it will take the same 10 chipmunks four times as long. This means it will take them 4×5 hours, or $\boxed{\text{20 hours}}$.

    Solution 2: Let r be the number of treehouses that one chipmunk can build in an hour, and let t be the amount of time it takes for ten chipmunks to build twenty treehouses. Since all chipmunks build treehouses at the same rate, r stays constant. Putting this information into an equation gives the result:

$$\textit{number of chipmunks} \times\ r \times\ \textit{time} = \textit{amount of treehouses built}$$

From this equation, r can be calculated by plugging in the information provided in the question:

$$2\ \textit{chipmunks} \times r \times 5\ \textit{hours} = 1\ \textit{treehouse}$$

Then, divide both sides by 10 to find that $r = \frac{1}{10}$ treehouses per hour. From here, if the number of chipmunks is set to 10, and the number of treehouses is set to 20, t can be found.

$$10\ \textit{chipmunks} \times \frac{1}{10} \times t\ \textit{hours} = 20\ \textit{treehouses}$$

Therefore, t, which is the amount of time it takes to build 20 treehouses with 10 chipmunks, is $\boxed{20 \text{ hours}}$.

12. In rectangle $ABCD$, AB is 1 less than twice the length of BC. If the perimeter is 22, what is the area of $ABCD$?

    Solution: First off, let AB be the length of the rectangle, and let BC be the width. Since the perimeter of a rectangle is equal to $2 \times (\textit{length} + \textit{width})$, this means the perimeter is also equal to $2 \times (AB + BC)$. The perimeter is 22, so $2 \times (AB + BC) = 22$. Then, divide both sides by 2 to get $AB + BC = 11$.

    Also, since AB is 1 less than twice the length of BC, $AB = 2 \times BC - 1$. Then, to solve for BC, substitute $2 \times BC - 1$ for AB in the first equation:

    $$\begin{aligned}(2 \times BC - 1) + BC &= 11\\ 3 \times BC &= 12\\ BC &= 4\end{aligned}$$

    Then AB must equal $2 \times BC - 1 = 2 \times 4 - 1 = 7$. Finally, the area must be $AB \times BC = 4 \times 7 = \boxed{28}$.

13. Matt is sleeping through his math final, which happens to be 100 questions long. When he is woken up with 15 minutes remaining, he starts, but cannot finish his test. If Matt had answered three-fifths of the questions on his math final, how much longer would he have needed? Express your answer in minutes.

    Solution: Since Matt answered three-fifths of the questions, he must be able to answer $100 \times \frac{3}{5} = 60$ questions in 15 minutes. This means he can answer $60 \div 15 = 4$ questions every minute.

In order to answer 100 questions, Matt must have needed $100 \div 4 = 25$ minutes to finish the test. Since he has already been working for 15 minutes, he needs $25 - 15 = \boxed{10 \text{ minutes}}$ longer.

14. The sum of the area and the perimeter of a rectangle with integer sides is 38. What is the largest possible area of the rectangle?

    Solution: Let a be the length of the rectangle, and let b be the width of the rectangle. Then $area + perimeter = ab + 2a + 2b = 38$. From this point, plug in integers for a, starting with $a = 1$.

    If $a = 1$, then $b + 2(1) + 2b = 38$, so $b = 12$, and the area will equal $1 \times 12 = 12$.

    If $a = 2$, $2b + 2(2) + 2b = 38$, so $4b = 34$, but b will not be an integer, so $a = 2$ is not a valid solution.

    If $a = 3$, then $3b + 2(3) + 2b = 38$, so $5b = 32$, but b will not be an integer, so $a = 3$ is also an invalid solution.

    If $a = 4$, then $4b + 2(4) + 2b = 38$, so $b = 5$, and the area will equal $4 \times 5 = 20$. If $a = 5$, then $b = 4$ because it will be the same rectangle as the $a = 4$ rectangle.

    From here, since $a > b$ when $a \geq 5$, b would have to be less than 5. Since b has to be a positive integer, b must be 1, 2, 3, or 4. However, every rectangle with side lengths 1, 2, 3, and 4 have been tested, so there is no need to test for values where $a \geq 5$.

    Therefore, since $20 > 12$, the largest possible area for the rectangle is $\boxed{20}$.

15. Michael and Jason have a pizza eating competition. The pizzas are perfectly shaped squares with side lengths that are whole numbers greater than 1. If the total area of one pizza Michael ate is 25 in^2 less than the total area of one pizza Jason ate, what is the side length of the larger pizza?

    Solution: Let the side length of Michael's pizzas be m, and let the side length of the Jason's pizzas be j. Since the area of one of Michael's pizza is 25 in^2 less than the area of one pizza Jason ate, this means $m^2 + 25 = j^2$, or that $m^2 + 5^2 = j^2$.

    Note that since this equation is in the form $a^2 + b^2 = c^2$, this problem is an application of the Pythagorean Theorem. Therefore, since m, 5, and j are integers, the numbers m, 5, and j make up a Pythagorean Triple. Logically, since the 5-12-13 Pythagorean Triples includes 5 as one of the two smaller numbers, this must be the solution! Therefore, m must equal 12 inches, and j must equal $\boxed{13 \text{ inches}}$.

16. Michael was very salty about losing the first eating competition against Jason, and called for a rematch. This time, they are eating extremely

salty square saltine crackers. Half of the area of each salty square saltine cracker is covered in pure salt. If Michael eats 24 in^2 more salty saltine crackers than Michael, how much more salt does he consume?

Solution: If Michael eats 24 in^2 more saltine crackers, then half of the 24 in^2 of crackers must be salt. Therefore, Michael ate $\frac{1}{2} \times 24 = \boxed{12 \text{ in}^2}$ more salt than Jason.

17. Jason is a math student who has trouble reading numbers- he flips the tens and ones digits of every number he sees. For example, he reads 24 as 42, and 63 as 36. If he calculated the area of a 51 by 12 rectangle, what is the positive difference between his answer and the correct area?

    Solution: The incorrect area that Jason calculated is $51 \times 12 = 612$. Next, since 51 and 12 are the incorrect numbers, the correct numbers must be 15 and 21, so the correct area is $21 \times 15 = 315$. The difference between these two areas is $612 - 315 = \boxed{297}$.

18. There are two rectangles with integer side lengths, both with area 48. What is the greatest possible difference between the perimeters of the two rectangles?

    Solution: First, list all possible rectangles with integer side lengths and an area of 48. If done correctly, 5 rectangles are given:

    1 by 48
    2 by 24
    3 by 16
    4 by 12
    6 by 8

    Then, in order to maximize the difference of the perimeters, the largest perimeter possible must be subtracted by the smallest perimeter possible. Clearly, the 1 by 48 rectangle will produce the largest perimeter, and the 6 by 8 rectangle will produce the smallest perimeter. The perimeter of the 1 by 48 rectangle is $2 \times (48 + 1) = 98$, the perimeter of the 6 by 8 rectangle is $2 \times (6 + 8) = 28$, so the maximum difference is $98 - 28 = \boxed{70}$.

19. Anh is trying to decide which books to read next, and so goes to her local bookstore. The bookstore has three sections: sci-fi, fantasy, and non-fiction, each with 10 books available. If she wants to choose two books from two different sections, how many ways can she choose them?

    Solution: Since there are three sections with ten books each, there are 30 possible first books that Anh could choose. Then, since her second book needs to be from one of the other two subjects, she has 20 possible choices for her second book.

This would mean there are $30 \times 20 = 600$ possible ways for her to choose them. However, this counts every possible scenario twice so 600 must be divided by 2 to get the final answer of $\boxed{\text{300 combinations}}$.

20. A, B, C, D, E, and F are digits from 1 to 9 and ABC and DEF represent 3 digit numbers. If $ABC + DEF = 1024$, what is the maximum possible value of $CBA + FED$?

    Solution: Since $ABC + DEF = 1024$, this means when C and F are added together, the ones digit must be 4. Since C and F are both one-digit numbers, $C + F = 4$ or $C + F = 14$. Since $CBA + FED$ needs to be as high as possible, $C + F$ should be set to equal 14.

    Next, since $C + F = 14$, the 1 gets 'carried over' to the addition of $B + E$, as shown in the addition problem below. This means $B + E + 1 = 2$ or 12, so $B + E = 1$ or 11. Again, since $CBA + FED$ needs to be as high as possible, so $B + E$ should be 11.

    Next, remembering to carry the one from the 11, $A + D + 1 = 10$, so $A + D = 9$.

    Finally, using place values, express CBA and FED using powers of 10:

$$CBA = 100 \times C + 10 \times B + 1 \times A$$
$$FED = 100 \times F + 10 \times E + 1 \times D$$

    Combining these two equations gives the result:

$$\begin{aligned} CBA + FED &= 100 \times (C + F) + 10 \times (B + E) + 1 \times (D + A) \\ &= 100 \times 14 + 10 \times 11 + 1 \times 9 \\ &= \boxed{1519} \end{aligned}$$

$$\begin{array}{r} {\scriptstyle 1\,1}\phantom{\mathrm{C}} \\ A\,B\,\mathrm{C} \\ +\,\mathrm{D}\,\mathrm{E}\,\mathrm{F} \\ \hline 1\,0\,2\,4 \end{array}$$

21. Twelve girls are trying to figure out how to split up into two teams of six for the All Girls Math Tournament. 3 of them are from Generic Middle School, 4 are from Standard Middle School, 2 are from Normal Middle School, and the other 3 are homeschooled. If each non-homeschooled girl must be on the same team as the other girls from her school, how many ways are there to split them up into two teams?

    Solution: The best way to solve this problem is to look at what two schools can be put in the same team (because one team must have two schools).

The first case is when Standard and Normal Middle School are on the same team. In this case, there is only 1 possible way to split the team because the remaining 6 girls are forced onto the other team.

Next, the second case is when Generic and Normal Middle School are on the same team. Since there are 5 spots filled by those schools, there is one spot left on their team. Since there are 3 homeschooled girls that can fill that spot, there are 3 possible ways to split the teams.

Next, the final case is when the Standard and Generic Middle School are on the same team, but this would put 7 girls on one team, so this is impossible.

Therefore, there are $1 + 3 = \boxed{4 \text{ ways}}$ to split them into two teams.

22. Aleena has a collection of 12 identical lavender-colored Sharpie markers, and she wants to distribute them all among three of her friends. If each friend must receive at least 3 Sharpie markers, how many ways are there to distribute the markers?

    Solution 1: First off, since each friend must get 3 Sharpies, then Aleena cannot choose where 9 of her Sharpies go to. Therefore, this question is the same thing as "How many ways can Aleena distribute 3 Sharpies among three friends?" Then, solve the problem by dividing it into cases:

    If Aleena gives all three Sharpies to one friend, there are three possible friends she can give all the Sharpies to, giving us 3 ways to distribute the pens in this case.

    If one friend gets two pens, and another gets one, there are 3 people who could get two pens, and 2 people remaining who could get one pen (because one person was already chosen). This gives us $3 \times 2 = 6$ ways to distribute the pens in this case.

    Finally, if each friend gets one pen, there is only one way to distribute the pens.

    Therefore, there are $3 + 6 + 1 = \boxed{10 \text{ ways}}$ to distribute the Sharpies.

    Solution 2: Since each friend must get 3 Sharpies, then Aleena cannot choose where 9 of her Sharpies go to. Therefore, this question is the same thing as "How many ways can Aleena distribute 3 Sharpies among three friends?"

    Using the balls and boxes principle, the Sharpies can be compared to the balls, and the friends can be compared to the boxes. Therefore, the number of possible combinations is $\binom{3+3-1}{3-1} = \binom{5}{2} = \frac{5\times4}{2} = \boxed{10 \text{ ways}}$ to distribute the Sharpies.

23. The first two terms of an arithmetic series are 5 and 10. The first two terms of a geometric series are also 5 and 10. What is the positive difference between the 5th terms of each series?

    Solution: Since an arithmetic sequence involves getting from one term to another by adding a constant d, d can be found by subtracting the first term from the second term. This means that d would be $10 - 5 = 5$. Then, the fifth term in the arithmetic sequence would be $5 + 5 + 5 + 5 + 5 = 25$.

    Next, since a geometric sequence involves getting from one term to another by multiplying by a constant r, r can be found by dividing the first term from the second term.This means that r would be $10/5 = 2$. Then, the fifth term in the arithmetic sequence would be $5 \times 2 \times 2 \times 2 \times 2 = 5 \times 16 = 80$.

    The difference between these two terms is $80 - 25 = \boxed{55}$.

24. Points A, B, C, and D are on a line, from left to right in that order. B is the midpoint of AC, and C is the midpoint of AD. If point E is not on the same line, what is the ratio of the area of $\triangle ABE$ to the area of $\triangle ADE$? Express your answer as a common fraction.

    Solution: Since B is the midpoint of AC, and C is the midpoint of AD, $2 \times AB = AC$, and $2 \times AC = AD$, so $4 \times AB = AD$.

    Next, because AD and AB are on the same line, $\triangle ADE$ and $\triangle ABE$ will have the same height, so the ratio of their areas is the same as the ratio of their bases. Finally, since $4 \times AB = AD$, the ratio of AB to AD is 1 to 4, or $\boxed{\frac{1}{4}}$.

25. The library is a very popular place in the small town of Bookworm. Every other day, Anika goes there work on her homework. Every third day, Ben goes there to read. Every fifth day, Cassandra goes there to meet up with the local book club. Given that all three were there on December 31 last year, how many days were *exactly* two of them there in January this year?

    Solution: Since Anika goes to the library every other day, and Ben goes to the library every third day, they will both be at the library every sixth day, so they will both be there on the 6th, 12th, 18th, 24th, and 30th.

    Likewise, since Cassandra goes to the library every fifth day, she will meet Anika there every tenth day, so they will meet on the 10th, the 20th, and the 30th.

    Finally, Ben and Cassandra will meet at the library every fifteenth day, so they will meet on the 15th and the 30th.

    However, since they were all at the library on the 30th, and the question asks for when *exactly* two of them were at the library, ignore the date of the 30th. Counting the remaining dates (6th, 10th, 12th, 15th, 18th, 20th, and 24th), there are $\boxed{\text{7 days}}$ where exactly two of them will be at the library.

26. Triangle ABC with non-zero area has $AB = 42$ and $BC = 73$. If AC has integer length, how many different possible values can it be?

    Solution: By the Triangle Inequality Theorem, the largest side of any triangle must be less than the sum of the other two sides. This means that $AC + 42 > 73$, or $AC > 31$, and $42 + 73 > AC$, or $115 > AC$. Since AC must be an integer, this means $32 \leq AC \leq 114$, which means there are 114-32+1 = $\boxed{\text{83 side lengths}}$ that AC could be.

27. The coordinates of a polygon are (0,0); (1,4); (5,0); (2,4); (2,2); and (5,2). What is the area of the polygon?

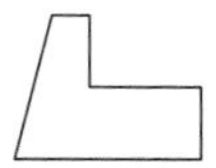

    Solution: This figure can be divided into two rectangles and a triangle, as shown below:

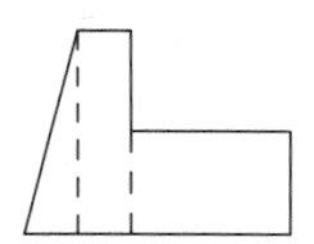

    The area of the triangle is $\frac{1}{2} \times 1 \times 4 = 2$. The area of the left rectangle is $1 \times 4 = 4$, and the area of the right rectangle is $3 \times 2 = 6$. This makes the total area $2 + 4 + 6 = \boxed{12}$.

28. Betty tosses a fair coin 6 times. What is the probability that she gets strictly more heads than tails?

    Solution: First up, since a coin has 2 possibilities for each flip, there are $2^6 = 64$ possible combinations for flipping a coin 6 times. Next, note that when there are 0, 1, or 2 tails, there are more heads flipped than tails flipped. Then, figure out the number of ways to get each case.

    If there are 0 tails, all coins are heads, so there is only 1 possible combination.

    If there is 1 tail, there are 6 coins that could be a tail, so there are 6 possible combinations.

    If there are 2 tails, there are 6 choices for the first tail, and 5 choices for the second tail, so there are $6 \times 5 = 30$ possible ways to pick two tails. However, this counts each scenario twice (because it counts picking the first penny, then the second, and vice versa) so divide 30 by 2 to get 15 possible ways to choose two tails.

Adding the number of ways in each case gives us $15+6+1 = 22$ ways to get more heads than tails. Finally, the probability that Betty gets more heads than tails is $\frac{\textit{ways desired result happens}}{\textit{total outcomes}} = \frac{22}{64} = \boxed{\frac{11}{32}}$.

29. Right triangle ABC has $AC = 8$ and $CB = 6$. M is the midpoint of AB. Pick point N on line CM with M between C and N such that $\angle CAB = \angle BAN$. Compute MN. Express your answer as a common fraction.

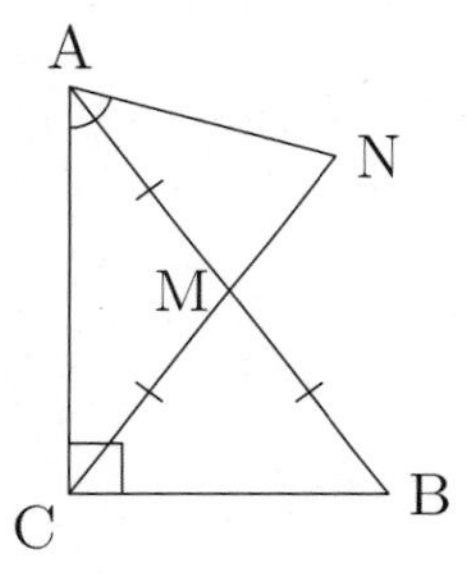

Solution: First, because of the Pythagorean Theorem, $AB^2 = AC^2 + BC^2 = 6^2 + 8^2 = 100$, so $AB = 10$. Since M is a midpoint, $AM = MB = 5$. Also, since MC is a median of a right triangle, $MC = MA = MB$, so $MC = 5$.

This means that $\triangle AMC$ is an isosceles triangle, so $\angle CAM = \angle ACM$. Furthermore, since $\angle CAM = \angle CAB = \angle BAN = \angle MAN$ and $\angle ACM = \angle ACN$, then $\angle ACN = \angle MAN$. Also, note that $\angle ANM = \angle CNA$. Since two pairs of corresponding angles in $\triangle CAN$ and $\triangle AMN$ have the same angle measure ($\angle ACN = \angle MAN$ and $\angle ANM = \angle CNA$), the two triangles must be similar.

Because the two triangles are similar, the ratio of the length of the corresponding side lengths will always be the same. This means

$$\frac{CA}{AM} = \frac{8}{5} = \frac{AN}{MN} = \frac{CN}{AN}$$

However, notice that $5 + MN = CN$. Substituting this in for CN results in a system of equations:

$$\frac{8}{5} = \frac{AN}{MN}$$

$$\frac{8}{5} = \frac{5+MN}{AN}$$

Using cross multiplication, realize that $5 \times AN = 8 \times MN$, or $AN = \frac{8}{5} \times MN$, and $5 \times (5 + MN) = 8 \times AN$. Using substitution, then it can be found that:

$$\begin{aligned} 5 \times (5 + MN) &= 8 \times (\frac{8}{5} \times MN) \\ 25 + 5 \times MN &= \frac{64}{5} \times MN \\ 25 &= \frac{64 - 25}{5} \times MN \\ \frac{25 \times 5}{39} = \frac{125}{39} &= MN \end{aligned}$$

Therefore, $MN = \boxed{\frac{125}{39}}$

30. Two circles, O_1 and O_2, of radius 6 and 8, respectively, intersect at points P and Q, as shown. AB is a line segment that passes through P and with one end on each circle. AQ is tangent to $\odot O_2$, and BQ is tangent to $\odot O_1$. Find the area of $\triangle AQB$.

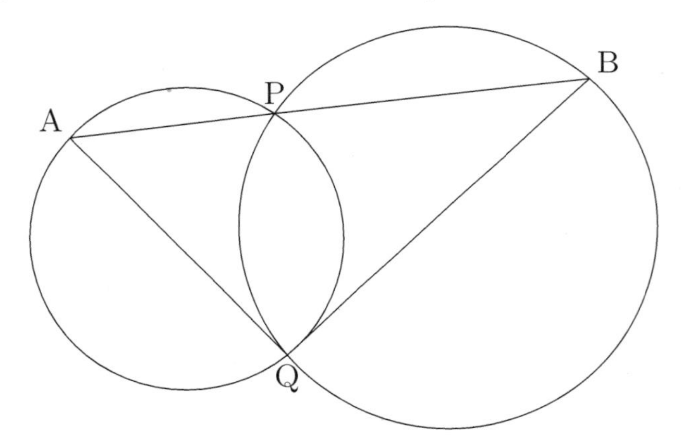

Solution: Using the fact that AB is a secant line that extends out of $\odot O_2$, and that AQ is a tangent line to $\odot O_2$, this means that

$$AB \times AP = AQ^2$$

Likewise, since AB is a secant line that extends out of $\odot O_1$, and that BQ is a tangent line to $\odot O_2$, this means that

$$AB \times PB = BQ^2$$

Adding these two equations gives us:

$$\begin{aligned} AB \times AP + AB \times PB &= AQ^2 + BQ^2 \\ AB \times (AP + PB) &= AQ^2 + BQ^2 \\ AB \times AB = AB^2 &= AQ^2 + BQ^2. \end{aligned}$$

By the Pythagorean Theorem, AQB is a right triangle, so $AQ \perp QB$. Furthermore, since QB is tangent to $\odot O_1$, this means $O_1Q \perp QB$. Since AQ and O_1Q are both perpendicular to QB, this means AQ and O_1Q are on the same line, so AQ must be a diameter of the $\odot O_1$.

Likewise, since AQ is tangent to $\odot O_2$, this means $O_2Q \perp AQ$. Since QB and O_1Q are both perpendicular to AQ, this means QB and O_2Q are on the same line, so QB must be a diameter of $\odot O_2$.

Then, since the radius of $\odot O_1$ is 6, AQ is 12, and since the radius of $\odot O_2$ is 8, QB is 16. Therefore, the area of the triangle is $\frac{1}{2} \times 12 \times 16 = \boxed{96}$.

Target Round

1. The area of a square is 12.25 in^2. What is the perimeter?

   Solution: First, note that the area of a square is equal to its side length squared ($Area = s^2$). Therefore, square rooting the area gets the length of one side: $\sqrt{12.25} = 3.5$. Then, multiply the side length by 4 to get the perimeter: $3.5 \times 4 = \boxed{14}$.

2. Bessie has 7 friends who are either cows or chickens. In total, her friends have 22 legs. How many of Bessie's friends are cows?

   Solution: Let x be the amount of cows Bessie has, and let y be the amount of chickens she has.

   Since she has 7 cows and chickens total, this means $x + y = 7$. Also, since her friends have 22 legs, and cows have 4 legs while chickens have 2 legs, $4x + 2y = 22$.

   Then, take $x + y = 7$. Subtracting x from both sides of the equation results in $y = 7 - x$. Then, substitute $7 - x$ for y in the second equation:

$$\begin{aligned} 4x + 2y &= 22 \\ 4x + 2 \times (7 - x) &= 22 \\ 4x + 14 - 2x &= 22 \\ 2x + 14 &= 22 \\ 2x &= 8 \\ x &= \boxed{4 \text{ cows}} \end{aligned}$$

3. Adam uses sticky notes to remind him to do his homework. He uses 1 sticky note on the first day of school, 2 sticky notes on the second day of school, and so on, using twice as many sticky notes each day. How many sticky notes does he use in total by the end of the sixth day of school?

   Solution: On the first day, he uses 1 sticky note, on the second day he uses 2, on the third he uses 4, on the fourth he uses 8, on the fifth he uses 16, and the sixth he uses 32. This makes the total amount of sticky notes equal to $1 + 2 + 4 + 8 + 16 + 32 = 3 + 12 + 48 = \boxed{63 \text{ sticky notes}}$.

4. Tom and Laura have a total of 17 apples. The number of apples that Tom has is divisible by 4. The number of apples that Laura has is divisible by 3. How many apples does Laura have?

Solution: First, notice that if they both have a total of 17 apples, then neither one of them can have more than 17 apples. Next, since Tom has a number of apples that is divisible by 4, he has either 0, 4, 8, 12, or 16 apples. Likes, since Laura has a number of apples that is divisible by 3, she has either 0, 3, 6, 9, 12, or 15 apples.

Upon testing, the only set of two numbers that adds to 17 is 8 and 9. Therefore, Tom has 8 apples and Laura has $\boxed{\text{9 apples}}$.

5. The perimeter of a regular pentagon is equivalent to the perimeter of an equilateral triangle. If the side length of the pentagon is 6, what is the side length of the triangle?

   Solution: First, calculate the perimeter of the pentagon. Since it is a regular pentagon, it has 5 sides, and each side has the same length. Since the side length of the pentagon is 6, its perimeter is $6 \times 5 = 30$. Then, since an equilateral triangle has 3 sides of equal length, and its perimeter must also be 30, its side length is $30 \div 3 = \boxed{10}$.

6. A "holey number" is defined as a whole number whose digits consist only of 0, 4, 6, 8, and 9 (digits may be repeated). How many 3-digit holey numbers are there?

   Solution: In order to find the number of possible "holey numbers," find the amount of numbers that can go in each digit. First, since there are 5 possible numbers, there are 5 possible choices for the ones digit. Likewise, there are also 5 possible choices for the tens digit. However, a 0 cannot start a number, so there are only 4 possible choices for a number in the hundreds digit. To find the solution to the problem, simply multiply the amount of numbers that can be placed in each digit. This results in $5 \times 5 \times 4 = \boxed{\text{100 ``holey numbers''}}$.

7. Anne the Ant is walking on a line. Each second, she flips a coin. If it lands heads up, she walks a step forwards. Otherwise, she stays where she is. What is the probability that, after 6 seconds, she is exactly 3 steps away from where she started? Express your answer as a common fraction.

   Solution: If Anne only moves three spaces, that means she flipped 3 heads out of 6 coins. To calculate the probability of this happening, divide the number of ways to roll 3 heads by the number of total ways to flip 6 coins.

   First, find the number of ways Anne can get 3 heads in 6 coin flips. One way to do this is to list out every possibility, where H is heads, and T is tails (HHHTTT, HHTHTT, HHTTHT, etc.). However, a more efficient way to solve this to notice that there are $\binom{6}{3}$ (6 choose 3) ways to get 3 heads out of 6 coins. This equals $\frac{6!}{3! \times 3!} = \frac{6 \times 5 \times 4}{3 \times 2 \times 1} = 20$, so there are 20 ways to get 3 heads when 6 coins are flipped.

Next, find the total possible number of ways to flip 6 coins. Since there are 6 coins, and 2 possibilities for each coin (heads or tails), there are $2^6 = 64$ total combinations for 6 coins. This means the probability is $\frac{20}{64} = \boxed{\frac{5}{16}}$.

8. A six-sided die is weighted so that each odd number is equally likely, and each even number is equally likely as well. If the expected value of one roll is $3\frac{2}{3}$, what is the probability of rolling a one?

   Solution: Let the probability to roll any one of the odd numbers be x, and let the probability to roll any even numbers be y. This means the probability of rolling an odd number is $3x$, and the probability of rolling an even number is $3y$. Then, since the probability for anything to be rolled is 1, and the only possibilities are to roll an odd or an even number, $3x + 3y = 1$.

   Next, note that the average roll of the dice equals the sum of each number on the die multiplied by its probability of getting rolled. Therefore, since the probability of rolling 1, 3, and 5 is x, and the probability of rolling y is 2, 4, and 6, the average roll equals $1x + 2y + 3x + 4y + 5x + 6y = 3\frac{2}{3}$ or $9x + 12y = \frac{11}{3}$.

   Then, divide the first equation by three to get $x + y = \frac{1}{3}$. Then, subtract x from both sides of the equation to get $y = \frac{1}{3} - x$. Then, substitute $\frac{1}{3} - x$ for y in the second equation:

$$\begin{aligned} 9x + 12y &= \frac{11}{3} \\ 9x + 12(\frac{1}{3} - x) &= \frac{11}{3} \\ 9x + \frac{12}{3} - 12x &= \frac{11}{3} \\ -3x = -\frac{1}{3}x &= \frac{1}{9} \end{aligned}$$

   Therefore, the probability of rolling a one is $\boxed{\frac{1}{9}}$.

5th-6th Grade 2017 Solutions
Sprint Round

---

1. Susie went to the movies and brought twenty dollars. Her friends doubled the amount of money she had. Then she spent \$6 on popcorn and \$3 on a soda. How much money did Susie have left?

   Solution: First, her friends doubled her money, which means they multiplied it by 2. This means she has $\$20 \times 2 = \$40$. Then, she spends \$6 on popcorn and \$3 on a soda, which means she is left with $\$40 - \$6 - \$3 = \boxed{\$31}$.

2. Jenna was making a birthday cake, so she bought 2 dozen eggs. If she had 5 eggs left over, how many eggs did she use to make the cake?

   Solution: First, note that a dozen is equal to 12. Since she bought 2 dozen eggs, she must have bought $2 \times 12 = 24$ eggs. Then, since there were 5 eggs left over, Jenna must have used $24 - 5 = \boxed{19 \text{ eggs}}$.

3. To participate for the All Girls Math Competition, Casey needs to buy some pencils and erasers. A pencil costs 49 cents and an eraser costs 99 cents. If Casey wants to buy 2 pencils and 2 erasers, how much money would Casey need (in cents)?

   Solution: First, calculate the cost of the erasers. To do this, multiply the number of pencils by the price of each pencil, which is $2 \times 99 = 198$ cents. Then, calculate the cost of the erasers by multiplying the number of erasers by the price of each eraser, which is $2 \times 49 = 98$ cents. Then, add the cost of the pencils and the erasers to get the total. This makes the total $198 + 98 = \boxed{296 \text{ cents}}$.

4. Prom King and Queen are being decided at the All Girls Math Tournament Annual Dance. Jane received 5 more votes than Karen's amount of votes tripled. If Jane received 23 votes, then how many votes did Karen receive?

   Solution: Let k be equal to the number of votes Karen received. Then, if Karen's votes are tripled (or multiplied by 3), and 5 more votes are added, then that equals the number of votes Jane received. This means Jane's votes are equal to $3k + 5$. Then, since Jane got 23 votes, $3k + 5 = 23$. Then, subtract 5 from both sides of the equation to get $3k = 18$. Finally, divide both sides by 3 to get $k = \boxed{6 \text{ votes}}$.

5. At the end of school, Ms. Johnson decides to hand out 30 cookies among her class. There are 25 students in her class. If 20% of her

class is allergic to cookies, how many cookies should each student get? (Answer with a fraction in lowest terms.)

Solution: First, since $20\% = \frac{20}{100} = \frac{1}{5}$, then $\frac{1}{5}$ of the students in her class is allergic to cookies. This means $\frac{1}{5} \times 25 = 5$ students are allergic to cookies, and $25 - 5 = 20$ students are not allergic to cookies. Then, if 30 cookies are to be divided evenly among 20 people, then each student gets $\frac{30}{20}$ cookies, which is $\boxed{\frac{3}{2} \text{ cookies}}$.

6. The apps Temple Run and Candy Crush take up a total of 360 MB of space on Alice's phone. If Temple Run is three times as big as Candy Crush, how large is Candy Crush? (Note: A MB, or megabyte, is a unit that measures storage on electronic devices.)

   Solution: Let the amount of space Temple Run take up equal t MB. Since Candy Crush takes up three times the space of Temple Run, it takes up $3 \times t$ MB. Then, both of the apps together take up 360 MB, so $t + 3t = 4t = 360$. Then, dividing both sides of the equation by 4 shows $t = 90$. Then, Candy Crush is equal to $3t$, so it takes up $3 \times 90 = \boxed{270 \text{ MB}}$.

7. In a magic square, the sum of the numbers in each vertical, horizontal, and diagonal row must add up to the same number. Also, each number from 1 to 9 (including 1 and 9) must be used, and they can only be used once. In the magic square below, what number belongs in the highlighted box?

2		4
(shaded)	5	
		8

   Solution: First, realize that the diagonal row from the top left to the bottom right is completed. Since the sum of the numbers in this row is $2 + 5 + 8 = 15$, each row must have numbers equal to 15. Then, since the rightmost vertical row has a 4 and an 8, the number in the right center box must be $15 - 4 - 8 = 3$. Finally, the middle row has a 5 and a 3, so the number in the shaded box is $15 - 3 - 5 = \boxed{7}$. Here is a completed version of the magic square:

2	9	4
7	5	3
6	1	8

8. One day, Ms. Chou decided to measure the height of her eleven students. If their heights are 52 in, 53 in, 49 in, 56 in, 52 in, 54 in, 55

in, 53 in, 57 in, 47 in, and 52 in, what is the difference between the median height and the mode height, in inches?

Solution: First, list the heights in order from smallest to largest, like so:

$$47, 49, 52, 52, 52, 53, 53, 54, 55, 56, 57$$

Next, realize that the mode height is the height that is the most common height. Since 52 inches is the height of 3 students, and no other height appears more frequently, this is the mode. Then, the median number is the number in the middle of the list. Since 53 inches is in the middle of the list, it is the median. Then, the difference between the median and mode is $53 - 52 = \boxed{1 \text{ inch}}$.

9. In a regular deck of 52 cards, what is the probability of drawing an Ace of any suit? (Answer with a fraction in lowest terms.)

   Solution: There are four Aces in a deck of cards, and there are 52 cards, so the probability of drawing an Ace is $\frac{4}{52} = \boxed{\frac{1}{13}}$

10. There are about 7 billion people in the world, and about 60% of them live in Asia. If $\frac{2}{3}$ of Asians do **NOT** live in China, about how many people live in China?

    Solution: Since $60\% = \frac{60}{100} = \frac{3}{5}$, then $\frac{3}{5}$ of the world's population lives in Asia. This means $\frac{3}{5} \times 7,000,000,000 = 4,200,000,000$ people live in Asia. Then, if $\frac{2}{3}$ of Asians do NOT live in China, then $1 - \frac{2}{3} = \frac{1}{3}$ of Asians do live in China. This means $\frac{1}{3} \times 4,200,000,000 = \boxed{1,400,000,000}$ or $\boxed{1.4 \text{ billion}}$ people live in China.

11. Billy Jean draws 5 squares side by side, as shown in the figure below. If the squares have side lengths of 1, 2, 3, 4, and 5, in that order, what is the area of the figure?

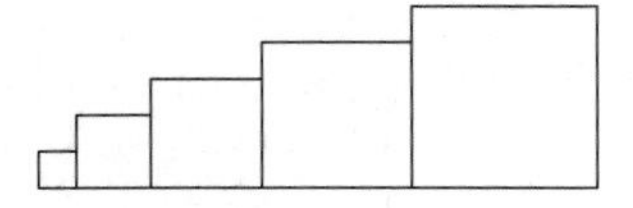

Solution: The area of the figure is the sum of the area of all 5 squares. To calculate the area of a square, multiply the side length of the square by itself. ($Area = side \times side$)

This means the area of the entire figure is equal to:

$$\begin{aligned}(1 \times 1) + (2 \times 2) + (3 \times 3) + (4 \times 4) + (5 \times 5) \\ = 1 + 4 + 9 + 16 + 25 \\ = \boxed{55}\end{aligned}$$

12. What is the perimeter of Billy Jean's figure?

    Solution: First, realize that the bottom side is equal to $1+2+3+4+5=15$, and the right side is equal to 5. That leaves the 5 tops of each squares, and the bold lines in the figure below. However, the sum of the tops of the square is also the sum of the side lengths of the squares, or $1+2+3+4+5=15$. Also, realize that the length of a bold line is the side length of one square minus the length of the square smaller than it. This means every bold line has a length of 1. This means the total perimeter is $15+5+15+(1\times 5)=\boxed{40}$.

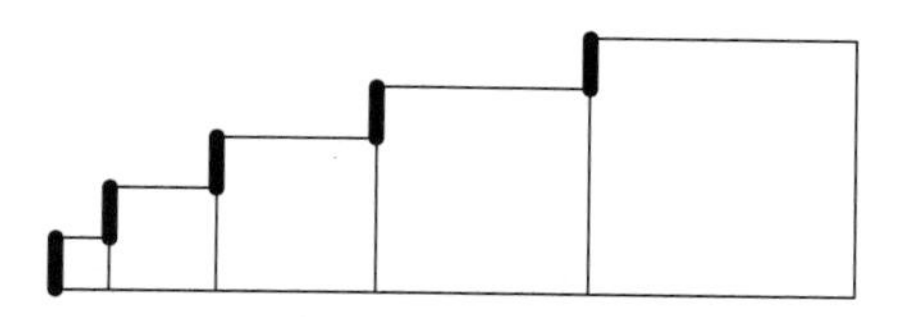

13. Matt and Linna were hungry one day, so they decided to have a chip eating-contest. Since Matt decides to be kind, he gives Linna a 30-second head start. Linna eats chips at a rate of 90 cm^2 per minute, and Matt eats at a rate of 8 chips per minute. If each chip is a triangle with a 5 cm base and 6 cm height, how many chips will Matt have eaten when he catches up to Linna?

    Solution 1: First, convert the rate Linna eats chips from cm^2 per minute to chips per minute. If a chip is a triangle with a 5 cm base and 6 cm height, one chip must be $1/2\times 5\times 6=15$ cm^2. This means she eats $90\div 15=6$ chips per minute.

    Then, since Linna gets a 30-second head start, or ½ minute start, she eats $1/2\times 6=3$ chips when Matt starts eating. Then, since Matt eats 8 chips per minute, he eats $8-6=2$ more chips than Linna does each minute. Then, since Linna starts 3 chips ahead, and Matt catches up at a rate of 2 chips per minute, it will take $\frac{3}{2}$ minutes for Matt to catch up with Linna.

    Finally, since Matt eats 8 chips per minute and has eaten chips for $\frac{3}{2}$ minutes, he must have eaten $8\times\frac{3}{2}=\boxed{12 \text{ chips}}$.

    Solution 2: First, convert the rate Linna eats chips from cm^2 per minute to chips per minute. If a chip is a triangle with a 5 cm base and 6 cm height, one chip must be $1/2\times 5\times 6=15$ cm^2. This means she eats $90\div 15=6$ chips per minute.

    Let t be the number of minutes that have passed from when Matt started eating to when Matt catches up to Linna. Since Matt eats 8 chips per minute, he has eaten $8t$ chips when he catches up to Linna.

    Then, since Linna eats 6 chips per minute, so she has eaten $6t$ chips since Matt started eating. However, since Linna gets a 30-second head start, or ½ minute start, she eats $1/2 \times 6 = 3$ chips before Matt starts eating. Therefore, she has eaten a total of $6t + 3$ chips.

When Matt catches up to Linna, they have eaten the same number of chips. This means $8t = 6t + 3$. Then, subtract $6t$ from both sides of the equation to get $2t = 3$. Finally, divide both sides by 2 to find that $t = \frac{3}{2}$ minutes.

Finally, since Matt eats 8 chips per minute and has eaten chips for $\frac{3}{2}$ minutes, he must have eaten $8 \times \frac{3}{2} = \boxed{12 \text{ chips}}$.

14. Linna was mad about losing the eating contest to Matt, so she decided to eat some cupcakes to make herself feel better. Each cupcake is 60 cm^3, and a quarter of the cupcake is frosting. If Linna ate 75 cm^3 of frosting, how many cupcakes did she eat?

    Solution: If ¼ of each cupcake is frosting, then there is $\frac{1}{4} \times 60 = 15$ cm^3 of frosting on each cupcake. Then, if Linna ate 75 cm^3 of frosting, she ate $75 \div 15 = \boxed{5 \text{ cupcakes}}$.

15. When Amy was hunting for Easter eggs, she found 30 eggs, each with 2 pieces of candy inside. She was really excited, so she immediately ate $\frac{1}{4}$ of her candy, and then hid the rest in her secret stash. However, her brother John found the secret stash and took an unknown amount of candy. Amy was sad that her brother stole her candy, so she ate $\frac{2}{3}$ of her remaining candy to cheer herself up. If there were 10 pieces of candy left afterwards, how many pieces of candy did her brother take?

    Solution: In order to find how many pieces of candy Amy's brother took, find the amount of candy before the brother stole some candy, and after the brother stole some candy.

    At first, she starts with $30 \times 2 = 60$ pieces of candy. She then eats a quarter of her candy, or $\frac{1}{4} \times 60 = 15$ pieces of her candy, leaving $60 - 15 = 45$ pieces remaining before her brother stole some.

    Then, at the end she is left with 10 pieces of candy. However, since Amy ate ⅔ of her remaining candy, the last 10 pieces must be $1 - \frac{2}{3} = \frac{1}{3}$ of her remaining candy. If 10 pieces is ⅓ of her remaining candy, she had $10 \div \frac{1}{3} = 10 \times 3 = 30$ pieces of candy after her brother stole his candy.

    This means he stole $45 - 30 = \boxed{15 \text{ pieces}}$ of candy.

16. Kevin has an obsession with pandas, so he decided to make a WALL OF PANDAS by purchasing lots of framed pictures of pandas. Each picture is 3 ft long and 2 ft tall. If Kevin's WALL OF PANDAS is 13 ft long and 11 ft tall, how many complete pictures can he possibly fit on his wall without rotating any of the pictures?

Solution: First, since the wall is 13 feet long, and each panda picture is 3 feet long, he can fit $13 \div 3 = 4^1/3$ pictures in one row. However, since there cannot be a part of a picture, only 4 pictures can fit in one row.

Next, since the wall is 13 feet tall, and each panda picture is 2 feet tall, he can fit $11 \div 2 = 5^1/2$ pictures in one column, so 5 complete pictures can fit in a column.

Since there are 4 pictures per row and 5 pictures per column, there can be $4 \times 5 = \boxed{\text{20 pictures}}$ on the wall.

17. Rosie is buying a new TV from TV Emporium, so she brought two coupons. The first coupon would give her a 20% discount, while the other coupon would give her a 10% discount and an additional $40 off. If each coupon would save her the same amount of money, how much would her new TV cost without using any coupons?

Solution: Let the price of the TV be x. Then, since $20\% = \frac{20}{100} = \frac{1}{5}$, the amount of money she would save with the first coupon is $\frac{1}{5}x$. Also, since $10\% = \frac{10}{100} = \frac{1}{10}$, the amount of money saved using the second coupon is $\frac{1}{10}x + 40$. Then, to find the price of the TV, set these two values equal to each other and solve for x:

$$\frac{1}{5}x = \frac{1}{10}x + 40$$
$$\frac{1}{10}x = 40$$
$$x = \boxed{\$400}$$

18. Casey draws five points on the same piece of paper. Then, she draws lines connecting each point to every other point. When she does this, how many lines would she have drawn? (Assume that no three points are on the same line.)

Solution: The easiest way to do this problem is to make 5 dots, and connect them all to one another, like the drawing below. If the lines in the figure are counted, there are a total of $\boxed{\text{10 lines}}$.

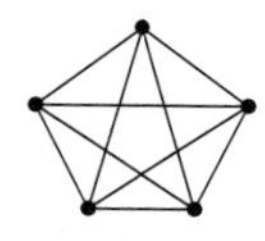

19. If ten people working for ten hours can complete ten jobs in total, then how many jobs can twenty people complete in twenty hours?

Solution: If 10 people working for 10 hours can complete 10 jobs in total, then $10 \times 2 = 20$ people working for 10 hours can complete

$10 \times 2 = 20$ jobs in total. Then, if 20 people working for 10 hours can complete 20 jobs, then 20 people working for $10 \times 2 = 20$ hours can complete $20 \times 2 = \boxed{40 \text{ jobs}}$ in total.

20. A circle is drawn inside a square so that the circle is touching the center of each side of the square. If the side length of the square is 6, what is the area of the circle? (Please leave answer in terms of π.)

    Solution: As shown by the figure below, the diameter is the same as the side length of the square, which is 6. Therefore, the radius is $6 \div 2 = 3$, and the area of the circle is $3^2 \times \pi = \boxed{9\pi}$.

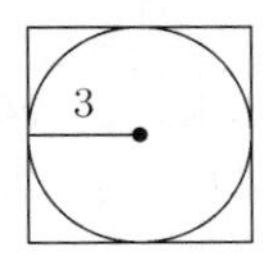

21. On Maddie's first four tests, she received a 96, an 88, an 84, and a 93. If Maddie is about to take a fifth test, and wants to get an average of 91, what score does she have to get on the final test?

    Solution: Let the score she needs to get on her final test be x. Then, realize that to calculate the average of the tests, take the sum of the test scores, and divide it be the number of tests. Since there are 5 tests, the average score is $(96 + 88 + 84 + 93 + x) \div 5 = 91$. Then, multiply both sides by 5 to get $361 + x = 455$. Finally, subtract 361 from both sides to get $x = 94$. Therefore, the score Maddie needs on her final test is a $\boxed{94\%}$.

22. When a bouncy ball drops from a certain height, it will bounce back up to only half of that height. If that ball is dropped from 1 meter high, after which bounce will the ball bounce to a height that is less than a centimeter?

    Solution: First, notice that after the first bounce, it will bounce back up $1/2$ a meter high, then after the second bounce, it will bounce to $1/4$ a meter high, and so on. This means that after the nth bounce, the ball will bounce $\frac{1}{2^n}$ meters high. Then, since 1 centimeter $= \frac{1}{100}$ of a meter, the answer must be the smallest number n such that $\frac{1}{2^n} < \frac{1}{100}$. $\frac{1}{2^6} = \frac{1}{64}$ and $\frac{1}{2^7} = \frac{1}{128} < \frac{1}{100}$, so the ball must bounce $\boxed{7 \text{ times}}$ before the ball bounces less than a centimeter high.

23. Sophia draws a house with a square body, a triangular roof, and a chimney in the shape of a trapezoid. Using the picture below, find the area of the house.

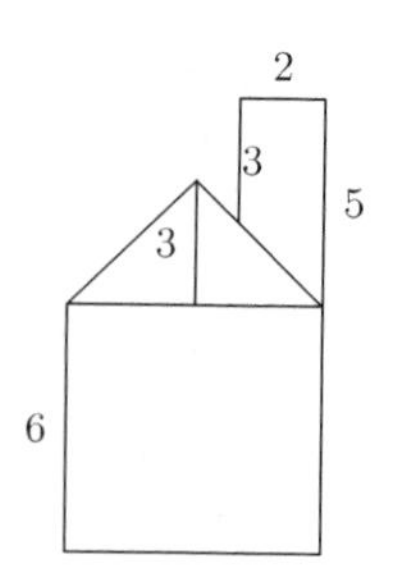

Solution: To calculate the area of the the house, calculate the area of the square, the triangle, and the trapezoid, and add them all together. The area of the square is $6 \times 6 = 36$. The area of the triangle roof is $1/2 \times 3 \times 6 = 9$. The bases of the trapezoid are 3 and 5, and the height is 2, so the $\frac{3+5}{2} \times 2 = 8$. Then, the area of the house is $36+9+8 = \boxed{53}$.

24. At the All Girls Car Wash, Anne and Bella are both washing cars. If Anne can wash a car in 30 minutes by herself and Bella can wash a car in 20 minutes by herself, how many minutes will it take them to wash a car together?

    Solution: If Anne can wash a car in 30 minutes, then she can wash 2 cars in an hour. If Bella can wash a car in 20 minutes, then she can wash 3 cars in an hour. Together, they can wash 5 cars in an hour, or 5 cars in 60 minutes, then they can finish washing one car is $60 \div 5 = \boxed{12 \text{ minutes}}$.

25. Andrea, Bethany, Caroline, and Daisy went to the movie theater together, and decided to sit in four seats in a row. If Andrea and Bethany are mad at each other and CANNOT sit next to one another, how many different ways can the four girls arrange themselves?

    Solution: In order to find how many ways they cannot sit next to each other, find the total number number of ways the girls can sit, and subtract the number of ways Andrea and Bethany DO sit next to each other. (Note: To help visualize this problem, it helps to draw a diagram with four 'seats', like the one below.)

    <u>Andrea</u>   <u>Bethany</u>   <u>Caroline</u>   <u>Daisy</u>

    If the four girls could sit anywhere, then any one of the four girls can sit in the first seat. Then, any of the three remaining girls could take the second seat. Then, one of the two girls left would take the third seat, and the final girl would sit in the fourth seat. Since 4 girls could sit in the first seat, 3 in the second, 2 in the third, and 1 in the last, there are $4 \times 3 \times 2 \times 1 = 24$ ways the girls can sit without restrictions.

    $$\underline{\quad 4 \quad} \times \underline{\quad 3 \quad} \times \underline{\quad 2 \quad} \times \underline{\quad 1 \quad} = 24$$

Then, if Andrea and Bethany want to sit together, treat them as one group of people. This would mean that there are three groups with three spots to choose from. Using the same logic from above, there are $3 \times 2 \times 1 = 6$ ways to arrange 3 groups in one row. However, since there are 2 possible ways Andrea and Bethany could sit together (Andrea could sit on the left, or Bethany could sit on the left), there are a total of $6 \times 2 = 12$ arrangements where the two girls are sitting together.

Then, subtract the number of ways they can sit together from the total number of ways the girls can sit to get $24 - 12 = \boxed{12 \text{ ways}}$.

26. Fifi was thirsty one day, so she bought three cartons of milk. Before she went to bed, she put the milk cartons in fridge so they wouldn't spoil. However, the fridge was broken, so each carton of milk has a $\frac{1}{3}$ chance of spoiling overnight. What is the probability that *none* of the cartons will spoil overnight?

    Solution: If the probability of one carton of milk spoiling is $\frac{1}{3}$, then the probability of the milk not spoiling is $1 - \frac{1}{3} = \frac{2}{3}$. If one milk carton has a $\frac{2}{3}$ chance of not spoiling, then the odds that all three milk cartons will not spoil are:

$$\frac{2}{3} \times \frac{2}{3} \times \frac{2}{3} = \boxed{\frac{8}{27}}$$

27. At 11:00 AM, Train Amnesty left the station, going north at a speed of 250 kilometers per hour. At 12:00 PM, the bullet train left the station, going north at a speed of 350 kilometers per hour. At what time will the bullet train catch up to Train Amnesty?

    Solution: If Train Amnesty leaves the station at 11:00 AM, it will have been traveling for a total of one hour. This means it has traveled a total of $250 \ \frac{miles}{hour} \times \ 1 \ hour = 250 \ miles.$

    Next, find out the rate at which the bullet train catches up to Train Amnesty. Since the bullet train travels at a speed of $350 \ \frac{miles}{hour}$, and Train Amnesty travels at $250 \ \frac{miles}{hour}$, the bullet train catches up at a rate of $350 \ \frac{miles}{hour} - 250 \ \frac{miles}{hour} = 100 \ \frac{miles}{hour}$.

    Then, if Train Amnesty is 250 miles ahead, and the bullet train catches up at $100 \, / \frac{miles}{hour}$, it will take $250 \ miles \div 100 \ \frac{miles}{hour} = 2.5 \ hours$, or 2 hours and 30 minutes, to catch up with Train Amnesty. Finally, since the bullet train left at 12:00 PM, and it takes 2 hours and 30 minutes to catch up, it will catch up at $\boxed{2{:}30 \text{ PM}}$.

28. Drew and Jenna have a deck of six cards numbered 1 through 6. Drew shuffles the deck, picks the top card, puts his card back in, and shuffles the deck. Jenna then picks a card, and they multiply their

two numbers together. If the result is odd, Drew wins the game, and if the result is even, Jenna wins the game. What is the probability Jenna wins the game? (Answer with a fraction in lowest terms.)

Solution: First, notice that when you multiply an odd number by an odd number, it will be odd. However, when you multiply an even number by an odd number, an odd number by an even number, or an even number by an even number, it will always be even. So, this means it's easier to find the odds of Drew winning.

Since there are 3 odd numbers between one and six inclusive, and there are 6 cards, Drew has a $\frac{3}{6} = \frac{1}{2}$ chance of drawing an odd number card. Using the same logic, Jenna will has a $\frac{1}{2}$ chance of drawing an odd card. This makes the probability of Drew winning $\frac{1}{2} \times \frac{1}{2} = \frac{1}{4}$.

If Drew has a $\frac{1}{4}$ chance of winning, then Jenna must have a $1 - \frac{1}{4} = \boxed{\frac{3}{4}}$ chance of winning.

29. Right now, Alexa is four times as old as her brother Michael. In four years, she'll be twice as old as Michael. How old will Alexa be 10 years from now?

Solution: Let Alexa's current age be a, and let Michael's current age be m. Since Alexa is four times as old as Michael, this means $4 \times m = a$.

Four years from now, Alexa's age will be $a + 4$, and Michael's age will be m + 4. Since Alexa will only be twice as old as Michael in four years, $2 \times (m + 4) = a + 4$. Simplifying this results in $2m + 8 = a + 4$.

Then, using substitution, $4m$ can be substituted for a in the equation $2m + 8 = a + 4$:

$$\begin{aligned} 2m + 8 &= a + 4 \\ 2m + 8 &= 4m + 4 \\ 4 &= 2m \\ m &= 2 \end{aligned}$$

This means that Michael is 2 years old today. Since Alexa is $4 \times m$ years old, she is now $4 \times 2 = 8$ years old today, and that 10 years from now, she will be $8 + 10 = \boxed{\text{18 years old}}$.

30. A mouse has a block of cheese that is in the shape of a triangular prism. The base of the block of cheese is in the shape of a right triangle with legs that are 6 inches long and 8 inches long. The height of the block is 4 inches. If the mouse ate $\frac{1}{3}$ of the cheese, how much cheese is left (in cubic inches)?

Solution: To find the volume of a prism, find the area of the base, and multiply it by the height. The base is a right triangle with legs that are 6 and 8 inches, so the area of the base is $1/2 \times 6 \times 8 = 24$ in^2. Next, multiply the area of the base to get the volume: $24 \times 4 = 96$ in^3. However, since the mouse ate ⅓ of the cheese, then he ate $\frac{1}{3} \times 96 = 32$ in^3 of cheese, leaving $96 - 32 = \boxed{64 \text{ in}^3}$ of cheese remaining.

Target Round

1. Jessica just turned 16, so she can finally buy her first car! She went to the car dealership and found two cars she really liked, both of which were on sale. The pink car was \$35,000, but the dealer took \$3,000 off the price, and then took 20% off the discounted price. The purple car was \$40,000, but the dealer made it 25% off of the original price, then discounted it by another \$2,000. What is the difference in price between the two cars?

   Solution: First, find the price of the pink car. If the price was \$35,000, but then the dealer took \$3,000 off of that price, the new price is $\$35,000 - \$3,000 = \$32,000$. Then, if he takes 20% off of the discounted price, she will pay $100\% - 20\% = 80\%$ of the discounted price. This means the pink car will cost her $80\% \times \$32,000 = \frac{80}{100} \times \$32,000 = \frac{4}{5} \times \$32,000 = \$25,600$.

   Next, find the price of the purple car. If the price was \$40,000, but he takes 25% off of the original price, she will pay $100\% - 25\% = 75\%$ of the original price. This means the purple car's new price is $75\% \times \$40,000 = \frac{75}{100} \times \$40,000 = \frac{3}{4} \times \$40,000 = \$30,000$. Then, taking off an additional \$2,000 will mean she will pay $\$30,000 - \$2,000 = \$28,000$ for the purple car.

   Finally, that makes the difference of the two prices $\$28,000 - \$25,600 = \boxed{\$2,400}$.

2. Find the area of the shaded region. (Leave your answer in terms of π.)

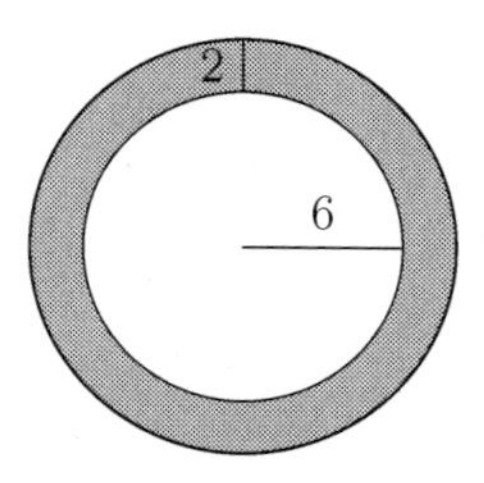

   Solution: In order to find the area of the shaded region, find the area of the big circle (which is the white and shaded areas combined) and subtract the small circle (which is just the white area). To find the area of these circles, use the area formula for a circle: $Area = radius^2 \times \pi$. The radius of the big circle is $6 + 2 = 8$, so the area of the big circle is $8^2 \times \pi = 64\pi$.

Next, find the area of the small circle. The radius of the small circle is 6, so the area is $6^2 \times \pi = 36\pi$. Then, subtract the area of the small circle from the big circle to get $64\pi - 36\pi = \boxed{28\pi}$

3. What is the area of the shaded region?

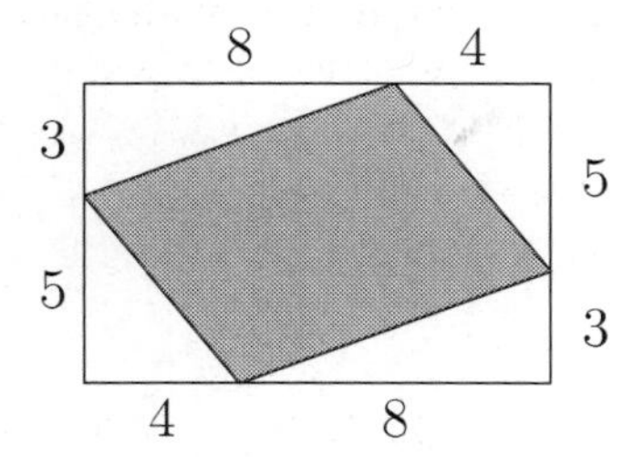

Solution: In order to find the area of the shaded region, first find the area of the big rectangle, and then subtract the area of the four right triangles around it.

First, find the area of the rectangle. The length is $3 + 5 = 8$, and the width is $8 + 4 = 12$, so the area is $8 \times 12 = 96$.

Next, find the area of the triangles. Two of the triangles have a base of 8 and a height of 3, so one of the triangle's areas is $1/2 \times 8 \times 3 = 12$. Then, the area of both triangles is $12 \times 2 = 24$.

The other two triangles have a base of 4 and a height of 5, so one of the triangle's areas is $1/2 \times 4 \times 5 = 10$. Then, the area of both of these triangles is $10 \times 2 = 20$.

This makes the area of the shaded region $96 - 24 - 20 = \boxed{52}$.

4. Candace and Stacy are playing a game. Candace picks a card from a deck of six cards numbered 1 through 6. Stacy picks a card from a deck of four cards numbered 1 through 4. Whoever has the higher number wins the game. What is the probability that Stacy will pick the higher number? (Answer with a fraction in lowest terms.)

Solution: The easiest way to solve this problem is count the number of ways Stacy can win, and divide that number by the total number of possibilities to get the probability.

First, find the number of ways Stacy wins. If Stacy draws a 4, she can win if Candace pulls a 3, 2, or 1, giving her three ways to win. If Stacy draws a 3, she can win if Candace pulls a 2, or 1, giving her two ways to win. If Stacy draws a 2, she can win if Candace pulls a 1, giving her one way to win. Stacy can't win if she draws a 1. This means she can win in $3 + 2 + 1 = 6$ ways.

Then, find the total number of cards they can draw. If Stacy can draw four cards, and Candace can draw 6 cards, they have a total of

$4 \times 6 = 24$ possible ways to draw cards. This makes the probability of Stacy winning $\frac{6}{24} = \boxed{\frac{1}{4}}$.

5. Joey was at the bookstore and wanted to buy some Harry Potter novels and some Percy Jackson novels. If she bought one book from each series, it would cost her \$35. If she bought all seven Harry Potter books and all five Percy Jackson books, it would cost her \$215. How much does one Harry Potter book cost? (Assume each book in a series is the same price.)

   Solution: Let h be the cost of one Harry Potter book and let p be the cost of one Harry Jackson book.

   Then, since one Harry Potter book and one Percy Jackson book cost \$35, $h + p = 35$. Also, since all seven Harry Potters and all five Percy Jacksons cost \$215, $7h + 5p = 215$.

   Then, take $h + p = 35$. Then, subtracting h from each side of the equation gets $p = 35 - h$. Then, substitute $35 - h$ for p into the second equation:

$$\begin{aligned} 7h + 5p &= 215 \\ 7h + 5 \times (35 - h) &= 215 \\ 7h + 175 - 5h &= 215 \\ 2h + 175 &= 215 \\ 2h &= 40 \\ h &= 20 \end{aligned}$$

   Therefore, one Harry Potter book costs $\boxed{\$20}$.

6. What is the number parallelograms in the figure below?

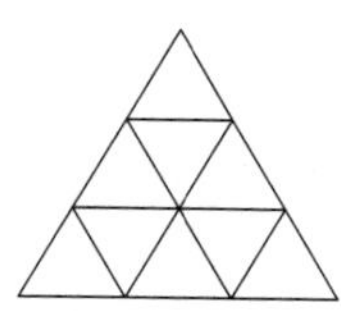

   Solution: There are $\boxed{\text{15 parallelograms}}$ in the figure below. (How many of each type of parallelogram are depicted below).

3 

3 

3

6 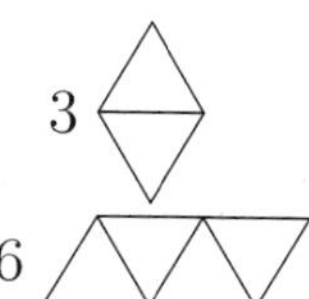

2 parallelograms with four triangles in bottom row, 2 in leftmost column, and 2 in rightmost column

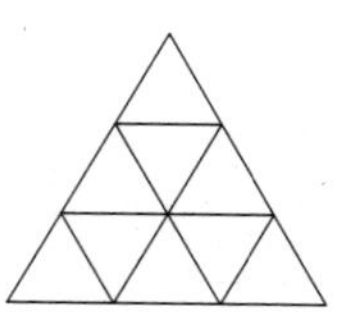

7. Elena has four numbered doors in front of her, each with a corresponding guard in front of it. She knows that one of the doors has treasure behind it, but she's does not know which door it is behind, so she decides to ask the guards some questions. However, EXACTLY one of these guards is lying. This is how they replied:

   Guard 1: The treasure is behind the door of the guard who is lying.

   Guard 2: Guard 3 is a liar.

   Guard 3: The treasure is behind Door #1.

   Guard 4: I am not a liar.

   Given their responses, which door is the treasure behind?

   Solution: The easiest way to do this is to go assume each guard is lying, and see if there is a contradiction if they are the liar.

   First, start with Guard 1. If he is lying, then the treasure is behind the door of someone who is telling the truth, so it can't be behind Door #1. However, if Guard 3 is telling the truth, then the treasure is behind Door #1. This is a contradiction, so Guard #1 must be telling the truth.

   Then, if Guard 2 is lying, then Guard 3 is telling the truth, so the treasure is behind Door #1. However, Guard 1, who also is telling the truth, says the treasure is behind the door of the guard who's lying, or Door #2. This is also a contradiction, so Guard 2 must be telling the truth.

   Then, if Guard 3 is lying, then the treasure is not behind Door #1. Then, Guard 2 would be telling the truth because he says Guard 3 is a liar, and Guard 4 would be telling the truth because he's not a liar. Guard 1 says the treasure is behind the door of the Guard who's lying, which is Door #3. There are no contradictions, so the treasure would be behind Door #3.

   Finally, if Guard 4 is lying, then Guard 3 is telling the truth, so the treasure is behind Door #1. However, Guard 1, who also is telling the truth, says the treasure is behind the door of the guard who's lying, or Door #4. This is also a contradiction, so Guard 4 must be telling the truth.

   Therefore, Guard 3 must be the liar, and the the treasure is behind $\boxed{\text{Door \#3}}$.

8. Michelle went outside to get her mail, but she realized that the mailman forgot to deliver her mail! Luckily, she sees that the mail truck

is 25 meters down the road, so she starts walking toward it at a speed of 2 m/s. However, the mail truck doesn't see her, so it drives away from her speed of 5 m/s, but has to take 10 seconds to stop at a house and deliver its mail. If each house's mailbox is 25 meters away from each other, and the mailman just finished delivering someone's mail, how long will it take for Michelle to catch up to the mail truck?

Solution: First, since the mail truck drives at 5 m/s for 25 meters, it drives for a total of $25 \div 5 = 5$ seconds. Then, since it stops for 10 seconds, and then continues its pattern, the mail truck moves a total of 25 meters every $5 + 10 = 15$ seconds.

Then, since Michelle walks at a constant rate of 2 m/s, she walks a total of $2 \times 15 = 30$ meters every 15 seconds. Since the mail truck moves 25 meters every 15 seconds, this means she gets $30 - 25 = 5$ meters closer to the mail truck every 15 seconds.

Finally, if she gets 5 meters closer every cycle, and she starts 25 meters away, it will take her $25 \div 5 = 5$ cycles to catch up to the mail truck. Then, since each cycle is 15 seconds, it will take $15 \times 5 = \boxed{75 \text{ seconds}}$ for her to catch up to the mail truck.

Made in the USA
San Bernardino, CA
15 March 2018